# WEARABLE TECHNOLOGY

www.royalcollins.com

# WEARABLE TECHNOLOGY

## *FROM BASICS TO ADVANCED APPLICATIONS*

**Kevin Chen**

*Books Beyond Boundaries*

ROYAL COLLINS

Wearable Technology: From Basics to Advanced Applications

Kevin Chen

First published in 2024 by Royal Collins Publishing Group Inc.
Groupe Publication Royal Collins Inc.
550-555 boul. René-Lévesque O Montréal (Québec) H2Z1B1 Canada

10 9 8 7 6 5 4 3 2 1



ISBN: 978-1-4878-1183-9

To find out more about our publications, please visit www.royalcollins.com.

# CONTENTS

# PREFACE

In 2012 Google surprised the world by announcing the arrival of Google Glass, a radical advancement in wearable technology. The wearable phenomenon has since grown rapidly, spreading to different technologies and applications with increasing market penetration. Underpinning this growth has been an Industrial Revolution, to a certain extent, with a growing industry based on hardware devices and software platforms, structured on novel business models and related ecosystems.

Wearable fitness and medical devices have perhaps seen the largest uptake to date. From Fitbit to Apple Health, these apps and wearables undoubtedly have great potential to improve the health and well-being of those wearing these devices. Many such wearables are based on traditional product-based business models. Makers of medical wearables have begun partnering with insurance companies to gain additional value from the data they collect.

The potential of wearable technology is significant. From advertising to tourism, medicine to gaming, there is hardly a sector that would not be impacted by this technology. Current applications of this technology include smartwatches, activity trackers, virtual reality (VR) devices, and medical applications. These are only the first fruits of a revolution that may lead to dramatic shifts in the way businesses interact with customers, the way in which a state relates to its citizens, and the way people interact with each other.

To date, there have been fewer applications in the fields of advertising, tourism, and e-commerce. This can be seen as the result of barriers in those business ecosystems—including the lack of standardized data collection, analysis, or customer feedback in these sectors. However, a significant portion of this book will examine potential future business models in these sectors, drawing on current market trends and scanning the horizon for future technology that may influence other business areas.

According to IDC, the global market growth of wearable technology is faster than that of any other segment of consumer electronics. The worldwide sales of wearables in 2014 were three times that of 2013, with over 192 million units sold. It is estimated that this figure may reach 1.26 billion units by 2019. Global sales may reach as high as USD 27.9 billion by 2019. In IDC's report on wearables for the first quarter of 2015, the total shipment volume was 11.4 million units, reflecting a 200% year-on-year increase and representing the eighth consecutive quarter of continuous growth.

According to the Global Wearable Devices Market Quarterly Tracking Report released by IDC, the global shipment of wearable devices reached 150 million units in the third quarter of 2023, an increase of 2.6% year-over-year. Despite the moderate growth, this still represents the highest shipment volume in the third quarter since 2021.

In the Chinese market, the China Wearable Devices Market Quarterly Tracking Report by IDC shows that the shipment volume of wearable devices in China reached 34.7 million units in the third quarter of 2023, up 7.5% year-over-year, indicating continuous growth in the overall market. Among these, the smartwatch market shipped 11.4 million units, up 5.5% year-over-year, with adult smartwatches accounting for 5.59 million units (up 3.9%) and children's smartwatches for 5.8 million units (up 7.2%). The wristband market shipped 3.98 million units, marking a 2.2% increase, while the ear-worn device market reached 19.24 million units, up 9.8%.

It is important to note that these reports only represent the tip of the iceberg in the wearable device industry, as the current market is predominantly focused on smartwatches, smart bands, and smart earphones. However, this does not fully exploit the potential of wearable devices.

Fundamentally, wearable devices are about sensor wear, with smartwatches, smart bands, smart glasses, smart rings, and smart clothing being just the initial product forms in the wearable device sector. Moving forward, as new types of sensors continue to emerge, a diversity of forms will likely define the major trend in the development of wearable devices. Some wearable devices might even seamlessly integrate into the user's body, becoming a natural part of the human anatomy unnoticeably.

Specifically, wearable devices can be divided into those worn on the surface of the body and those implanted within the body, which are wearable products worn outside the human skin, and implantable wearable devices placed inside the body.

Exterior wearable devices are the products we are more familiar with, primarily including smartwatches and smart bands. However, smartwatches and bands do not represent the entirety of wearable devices; they are merely one form of wearable technology worn outside the body. From the perspective of the entire wearable device industry, smartwatches, and bands, despite being early starters, occupy a relatively smaller segment of the market; the market potential for yet-to-be-exploited exterior wearable devices such as smart glasses, smart clothing, smart shoes, smart accessories, and smart underwear, is significantly larger than that of smartwatches and bands.

If we assume that the market size for smart glasses will be as large as that for smart bands and watches; that smart shoes will have a market three times the size of smartwatches and bands; that smart clothing will also have a market three times the size of smartwatches and bands; and that smart accessories will have a market size comparable to smartwatches and bands, excluding implantable wearable devices, medical wearable devices, the future market where smartphones become wearable phones, and smart underwear, the resulting figures are staggeringly large. It can be said that the market capacity for wearable devices far exceeds our current understanding.

In this book, I will look into the trajectory of wearable technology against the backdrop of global developments and the many popular applications it offers. This will build on real-life case studies and future projections as I analyze and explore the business models of wearable technology. I hope this book will help businesses, startups, and individuals interested in this industry find a good entry point and, in doing so, receive better returns on investments. Given the fast pace of development in this sector, this cannot be an exhaustive review of potential products or business models but is focused on areas that have been selected to represent the entire wearable technology industry.

The future of wearables should not be underestimated; it is not merely about our current focus on smartwatches and smart bands. It represents a digital twin of our planet and is a core carrier for the realization of the metaverse. Now, starting with this book, let us truly understand wearable devices and their real value.

# PROLOGUE

*We are right at the beginning of a technology revolution. People assume that the technology of tomorrow will be similar to that of today. What they do not know is that technology is exploding right in front of our eyes, and this will change everything.*

—LEE SILVER, PROFESSOR OF GENETICS AT PRINCETON UNIVERSITY

Over the past seventy years, the world has experienced an unprecedented revolution in information technology, shifting from the world economy of the industrial era to an Internet-based new global economy. The number of inventions and technological innovations in the past 20 years exceeds the entirety of the previous two to three centuries. This wave of technology has swept over our lives, transforming livelihoods and lifestyles.

At the time of writing, the rise of the smartphone era is driving what may be seen as the wearable device age, with information collection and presentation based on a smart end device closely associated with the human body. These devices will be incorporated into our body in a natural way, and blend into and rebuild our lives. Future economic development is likely to show a very different pattern. When the people at the center of the economy are entirely "hijacked," it is inevitable that the world economy will need to go through a radical transformation in which intelligent technology centered around wearables will lead the way.

Following the introduction of the Internet, we have seen significant changes in the way we live our lives and conduct our business. Looking ahead, wearable technology will build on this—causing a fundamental change in livelihoods, lifestyles, business-customer interactions, public management of services, and the working of the economy itself.

In the past 20 to 30 years, in the study of information networks and in bioscience, there have been some advances that may seem, at first glance, to be rather insignificant compared to advances like the discovery of the Higgs boson. However, what seems like pebbles dropped into the ocean of knowledge are causing ripples, which could bring about major waves in the near future. Without knowing much of the bigger picture, we have captured many of the opportunities presented at every point of technological development.

We have been amazed by advancements in bioscience over the years. Consider how our understanding of fertility has advanced. On July 25, 1978, Louise Brown was born in England, the first human baby born after conception by in vitro fertilization (IVF). The news of the birth sent shock waves throughout the world. Many people were worried, with words like "Frankenstein" used to describe this advancement. And yet Louise Brown grew up healthily and is happily married with her own children now. Two decades later, in 1997, a sheep named Dolly was successfully cloned by Ian Wilmut, a British embryologist. The birth of Dolly caused controversy, as it suggested the possibility of advanced mammals, including humans, being cloned. Neither the invention of IVF nor the cloning of Dolly the sheep have led yet to radical change in the way the world lives—but when these advances are coupled with smart technology, new doorways may open to treatments or changes in the way food systems work.

Smartphones have fundamentally changed our social habits, transforming our lives in many ways. The way we shop, what we read, and how we communicate have all been affected. Our working lives have been changed. The new technologies offer ways to make things easier or work more efficiently. Mobile phones are no longer tools for communication alone. Instead, they have changed our lives, our entertainment consoles, and our supermarkets.

The upcoming wearable age will continue to change modern life as we know it, gradually replacing the smartphone to further overturn the old ways of life.

Humans will enter what may be seen as a new epoch—the true age of "Big Intelligence."

Unlike smartphones, wearable devices will totally free our hands. Human-machine interaction will shift gradually to voice interaction or even brain wave interaction through subconscious thoughts. If you consider yourself "tied down" today by a smartphone that is the center of your life, the emergence of wearables offers the potential to change this. Wearables could release us from the black hole information paradox and make digital information more human-centered and human-serving through advanced communication technology. The wearables of tomorrow will not only act as your personal assistant but could also be your private nurse, monitoring your heart and blood pressure and contacting your doctor when there is an emergency.

Through DNA testing, today's medical profession is moving toward precision medicine. The aim is to fundamentally decode human diseases and predict potential health problems. With the help of wearables, both doctors and patients would have access to better data about their movement and other measures of health, allowing for the development of better, more effective medical treatments.

Biotechnology offers the potential to treat many diseases that are now incurable, including some cancers, Alzheimer's disease, and more. Advances in 3D printing technology could enable the printing of human organs on demand to replace failing ones, enabling increased longevity and removing the need for organ donors. Likewise, human cells could be printed in order to replace and eliminate cancer cells.

It is foreseeable that in the era of smart wearables, the entire value system, governance system, business system, and economic system of life will undergo fundamental changes. So, what will the era of wearables be like? The first time I published this book was in 2016. At that time, the concept of digital twins did not exist, nor did the concept of the metaverse. At that historical stage, we referred to the era formed by intelligent terminals and the intelligent wearabilization of all things as the era of smart wearables.

But now, we can express the era of smart wearables more clearly with the current vocabulary, which is the era of the metaverse and digital twins that we discuss and understand today. However, the current discussion and definition still have significant limitations; this is just a conceptual expression of the future under our current technology, culture, and cognitive environment. When we truly enter the era of smart wearables and the era of the metaverse, these concepts will definitely be revised by the technology, culture, and cognition of that future era. Therefore, in the foreseeable future, we can make some projections based on the present.

## THE EIGHT KEY SMART FUTURE TRENDS

The future we face will be an intelligent one, encompassing not only the smart products, smart home appliances, and autonomous driving concepts we discuss today, but also the application concepts of artificial intelligence (AI) like ChatGPT that are currently under discussion. Instead, it will be an era of interconnectivity and interaction built by intelligent terminals, namely smart wearable devices. Therefore, if we can look ahead and grasp these technological trends, we can not only see through the future but even proactively build a wonderful future life for ourselves.

After the personal computer was introduced in 1970, it, along with the Internet and bio-genetic engineering, sparked the second global technological revolution in just 40 years. With further enhancements in computing power, especially the emergence of NVIDIA GPUs, there was a significant improvement in computational abilities, enabling unprecedented advancements in the training of AI. This facilitated the creation of generative large language models like ChatGPT. It can be said that current advancements in AI are bringing about a new wave of transformation in human society. Clearly, the AI-driven transformation will have a deeper impact than previous technological revolutions sparked by the Internet and bio-genetics. For instance, AI drug discovery company Insilico Medicine, with the aid of AI technology, reduced the traditional ten-year cycle for new drug development to just 18 months. This is the miracle sparked by the combination of AI with biology and chemistry, a new technological revolution facilitated by AI. We firmly believe that human society will step into a future era dominated by intelligent technology in economic activities and social development, which we can call the "Smart Future" or the era of the metaverse, characterized by the following eight features:

### Mobile Economy

According to the Global Mobile Economy Report 2023 released by GSMA Intelligence in 2023, the report indicated that by the end of 2022, the number of unique mobile users worldwide reached 5.4 billion, with 4.4 billion being mobile Internet users. The usage gap for mobile Internet users has significantly narrowed over the past five years, decreasing from 50% in 2017 to 41% in 2022. However, there are still two main issues: the substantial disparity in mobile Internet usage among different countries and users, and the significant potential for enhancement in terminal intelligence and mobility.

The report highlights that, based on the still-not-fully-popularized smart wearable industry or the current foundation of mobile communication devices, in 2022, mobile technology and services generated 5% of global GDP, contributing USD 5.2 trillion in economic value added and supporting 28 million jobs within the broader mobile ecosystem. The report predicts that by 2030, the number of unique mobile users globally will increase to 6.3 billion and mobile Internet users to 5.5 billion. At that time, the global connection share of 4G will decrease from 60% in 2022 to 36%, while 5G connections will increase from 12% in 2022 to 54%. Licensed cellular Internet of Things (IoT) connections are expected to grow from 2.5 billion in 2022 to 5.3 billion. Moreover, the contribution of the mobile industry to global GDP is projected to rise from USD 5.2 trillion in 2022 to over USD 6 trillion by 2030.

Clearly, we are gradually transitioning from the traditional Internet era, which was PC-centric, to the mobile Internet era, characterized by mobile devices such as smartphones, VR, and wearable devices. This transition heralds a new mobile economic model: the entire market economy will shift from fixed networks to consumers' mobile devices. Furthermore, with the introduction of Starlink technology, humanity will move from the current limited mobility to an era of unlimited mobile communication that integrates space, air, and sea.

As the intelligence of terminal products deepens, and with AI chips and Starlink communication as the foundation, intelligent terminal products will exhibit intelligent communication and decision-making from end to end. It is foreseeable that in the next 20 years, human society will enter an era that relies on algorithms and is constructed and governed by them. In the world of algorithms, wearable technology will become an indispensable intelligent organ in human life.

## Sharing Economy

Since the emergence of the sharing economy model, represented by Uber, the sharing economy, with the aid of Internet technology, is profoundly transforming the ways people work, conduct business, and live. Especially under the impact of the pandemic, the development trend of the sharing economy has accelerated. People no longer need fixed offices or contracts without specified job duties and enjoy flexible working hours and considerable income. It's not just changing how people live; it's also transforming how they work. By efficiently allocating idle resources to maximize benefits, the greatest appeal of the sharing economy lies in its flexibility: almost anyone can participate and benefit from it at any time.

An Uber driver can be a university professor or a white-collar worker spending most of the day working behind a desk. Regardless of your routine job, as long as you have spare time and a vehicle that meets its requirements, you can be an Uber driver ready to earn some extra money. Research shows that Uber drivers' average education level is rather high. Nearly half have bachelor's degrees and above, a much higher proportion than taxi drivers (18%) and above the average of the entire labor force (41%).

Instacart is a similar example. It gives the opportunity to earn money while shopping. As long as you have a smartphone, are over 18 years old, and are able to lift heavier items, you can be part of Instacart, working at your own convenience. When you are shopping in a supermarket, and you see a neighbor has placed an order on Instacart, you can take the order, shop for the person, and deliver it to their door. The pay is about USD 25 per hour.

Other players include Zipcar (car sharing), Airbnb (house sharing), BookCrossing (book sharing), and others, all providing platforms that enable consumers to rapidly identify the commodities they need at a much lower price.

This model is the result of the unique data flow generated by mobile technology that happens anywhere at any time.

As autonomous driving technology advances further, once cars possess true autonomous driving capabilities, car owners, after fulfilling their commuting needs, can deploy their vehicles for commercial operations while working in the office. They can earn money by working while simultaneously allowing their self-driving cars to provide transportation services for others, generating additional income. When we go home to sleep in the evening, autonomous vehicles can continue to work, helping us earn money. This exemplifies the charm of the sharing economy.

### Boundless Computing

Under the impetus of AI technology, China and the world are stepping into an era centered on cloud computing, a comprehensive computational power era. Cloud computing centers are responsible for data convergence and processing, handling the transmission, storage, and computation of massive and complex information data. In the future, the information carriers we encounter will often be various screens, such as smartphones, tablets, and TV walls that are even thinner and lighter than today's devices, or even virtual holographic screens. In essence, screens will become ubiquitous. While screens are merely the carriers for information presentation in our lives, all information processing will be undertaken by unseen computational power.

Boundless computing will change the way we work as well as our lifestyles. Traditionally, the average work day has been from nine to five, but as times change, we will be presented with numerous job opportunities outside these traditional workplaces and hours powered by the mobile network of information. In the US, only 35% of employees need to be physically present at the office during work hours. The rest of the workforce is already able to work at home or other venues, engaging in innovative businesses or providing customer services. Mobile smart devices enable instant communication with employers. The mobile network brings businesses and working environments together as never before.

Furthermore, the era of the metaverse, built on smart wearables, will trigger a new form of international division of labor. This time, the core of the division will not be centered on physical production and manufacturing but will revolve around the international division of labor concerning a series of information-processing tasks sparked by wearable devices. In the future, Hong Kong (China) may be the global data center for ophthalmology and the US the center of pediatric data processing. Boundless computing will overcome geographic barriers and boost data processing platforms established around the world with countries leveraging their unique advantages in industrial data processing to serve the needs of the entire world.

## AI

In 2011, the supercomputer Watson competed on the US game show *Jeopardy!* against former champions, ultimately winning first place. In Japan, a female-looking humanoid robot dressed in a kimono known as Aiko Chihira appeared at the information desk in the Mitsukoshi department store in Tokyo to provide customer services and introduce store events. This humanoid robot is developed by Toshiba to perform as a receptionist.

The advancement of AI will have a profound impact on work and daily life. Particularly in the field of industrial manufacturing, intelligent robots or the use of mechanical arms may increase productivity considerably. Automated self-service is now readily available in many gas stations and supermarkets around the world. This overwhelming trend of process automation is gradually replacing human labor with robotics in all aspects of business life. This will undoubtedly pose a threat to human employment prospects.

In April 2022, Academician Pu Muming from the Chinese Academy of Sciences shared his insights in *Future China* regarding when machines could replace human jobs, mentioning that AI is expected to replace 90% of human jobs by 2075.

On June 14, 2023, the consulting firm McKinsey released a research report titled "The Economic Potential of Generative AI." In the report, analysts explored the impact of AI's exponential development on the global economy through a study encompassing over 80% of the global working population. They examined which industries would be most affected and who would face unemployment threats and concluded that the timeline for AI to replace human jobs had been significantly moved up by a decade. Between 2030 and 2060 (with a midpoint in 2045), 50% of jobs will gradually be replaced by AI.

I have also expressed publicly on multiple occasions that in the near future, all regular and rule-based jobs in human society will be replaced by AI. This will accelerate our entry into a new era of human-machine collaboration.

However, this is just the beginning. When AI integrates with intelligent wearable technology, it will not only smartly become our life assistant but also be omnipresent. At the same time, with the development of generative AI technologies capable of self-learning, there could be potential crises influencing and dominating our entire human society. Nonetheless, the trend of AI is unstoppable …

**Smart Living**

Let us visualize a future lifestyle like this.

When you get in your car after a long day of work, your wearable device analyzes your energy level, mood, and the day's workload, understands your preferences at this moment, and automatically plays the perfect music for what you need, just as you start the engine. At the same time, it informs smart appliances at home and humanoid robot butlers to prepare the house for the right temperature and lighting. Your bath water will be ready if you want it, and your healthy dinner will be prepared.

You don't need to switch anything on when you arrive at home. Your smart home appliances know your habits and will prepare these options without your active input, using the data feed from your wearables. The choices only appear when you want them.

No human butler can provide a better level of service. Servants may get to know your food and clothing preferences over the years, but knowing what goes on in your mind is virtually impossible. Once wearable devices and brain-computer interface (BCI) technologies are integrated, humanoid robots will assist us in accomplishing all aspects of intelligent management. The future life will be crafted for us by the collaborative efforts of numerous intelligent devices and AI that "understand" us. This kind of smart living is steadily approaching everyone.

## Regenerative Medicine

Regenerative medicine is a cutting-edge medical technique to repair human organs or tissue using stem cells. Growing tissues and organs in a laboratory and implanting them stimulates the body's own repairing mechanism to functionally heal damaged or irreparable organs or tissues in order to restore or establish normal function.

In 2006, Professor Shinya Yamanaka from Kyoto University and his team generated induced pluripotent stem cells (iPS cells). He was awarded the 2012 Nobel Prize in Physiology and Medicine. iPS cells could be used to grow all active human cells and tissues for teeth, nerves, retinas, cardiomyocytes, blood cells, and liver cells, which could be implanted to restore damaged organs or tissues. Doctors could use this technology to grow stem cells from the patient's own cells, in order to discover the way a particular disease is developing (the "pathogenesis"). Targeted treatment could then be developed to address that pathogenesis at the cellular level.

Regenerative medicine will become easier and more effective in the future. With the progress of bioprinting, 3D printing technology could be used to directly "print" human organs in the future. And 3D-printed nano-robots would serve as a safeguard inside human bodies, patrolling around the clock to identify cancer cells. They could even use a self-triggered shape-shifting function to treat or kill cancerous cells.

In a word, humans would no longer need to be afraid of losing the battle against a range of diseases. Medicine will likely be a leading sector in embracing the advances offered by smart technology, particularly with the arrival of the age of wearables, and will be radically changed as a result.

## Thought Sharing

In the Bollywood movie *PK*, the alien PK doesn't communicate through spoken language or facial expression but rather uses handshakes and telepathy. This way of communication makes lying very difficult, but learning from one another is quite simple.

Stephen Hawking, one of the greatest minds in recent history, developed the symptoms of amyotrophic lateral sclerosis (ALS) when he was 21. As the disease progressed, he became wheelchair-bound, only able to move his two eyes and three fingers. By the end of his life, he employed a novel device that used facial expressions to generate speech, adding to his expressiveness. In fact, these technologies are not science fiction, but BCI technologies that are being realized.

Many people might be enthusiastic about sharing their thoughts or becoming telepathic. We can all recall moments during exams when we wish we had a photographic memory. When Ericsson conducted a global survey among smartphone users, 40% of them expressed the wish "to be able to communicate with others using thoughts via wearables." Two-thirds of the participants even believed that such communication would be quite common in the future.

Indeed, thought sharing is on its way to becoming a reality, particularly through the burgeoning technology of BCIs. Strictly speaking, BCI technology is a type of wearable device, one that reads and writes brainwaves and consciousness, serving as a brain-centric intelligent wearable technology. With the integration of BCI and AI technologies, humanity can overcome the barriers of language communication. For individuals with language impairments or those who face communication challenges due to physiological reasons, the combination of BCIs and AI-powered speech generation technologies can accurately interpret and articulate their thoughts using the brain's intentions, read by BCIs, and then vocalized through an AI speech system.

Furthermore, the barriers to cross-linguistic communication will be eliminated in the era of intelligent wearable devices like BCIs. Communication in any language can be presented in the interlocutors' native tongues directly to their brains, and expressions can be articulated in the other party's language to facilitate conversation. The concept of thought sharing is inching closer to reality with the advent of BCI technology.

**Forecast Monitoring**

Data generated by search engines, GPS receivers, social media, and other systems will increasingly impact the entire business world through the advancement of data analysis technology. Google, Facebook, and Twitter activity are used together with smartphones to provide extensive consumer behavior data. Big data analytics build consumer profiles to precisely predict the shopping behavior of individuals. Supported by GPS, GIS, and SLAM (simultaneous localization and mapping) systems in iOS or Android platforms, computers can also predict user movement by tracking the use of smartphones. Visual sensor systems work to monitor our daily activities around the clock, and privacy is pushed to the minimum. For that reason, futurist Patrick Tucker believes we will be living in a "naked future" where everything is monitored, and privacy will no longer exist.

Recent research has indicated that the Starlink network could replace GPS positioning services, offering accuracy that exceeds civilian GPS by tenfold and is less susceptible to interference. According to calculations by Todd Humphreys from the University of Texas at Austin, this low earth orbit positioning system using real-time orbital and clock data from Starlink could allow for user location accuracy within 70 centimeters. This is about ten times more accurate than the GPS systems currently widely used in smartphones, watches, and cars. Moreover, utilizing the Starlink network for this purpose would not consume more than 1% of its downlink bandwidth or exceed 0.5% in power consumption. Consequently, futurist Patrick Tucker believes that we will live in a "Naked Future," where surveillance is ubiquitous, and privacy is virtually nonexistent.

In the era of wearable devices, all data will become more precise, and commercial activities will be executed in mere seconds. Those who possess more effective data will be able to offer personalized services to customers, thereby gaining their attention and trust. Evidently, data privacy will become one of the foremost issues in the future. Countries worldwide will grapple with the massive data influx generated by intelligent wearables, and debates surrounding human data privacy will emerge as a focal point and contentious topic in the new democratic discourse.

## CONSUMER MODELS IN THE FUTURE

### Mainstreaming the Limited Personalized Consumer Model

In the world of commercial consumption, the truth is not what matters; what's crucial is convincing consumers to believe that what they're presented with is the truth. Clearly, each era has its consumption trends and patterns, and in the 21st century, we are facing a mainstream shift toward personalized consumption. However, this personalization is somewhat limited; it's more about providing consumers with more options, as the array of products available has expanded significantly.

The launch of the iPhone product instantly reduced the distance between businesses and users, the first of a generation of increasingly advanced smartphones. All the commodities in the world are now at the consumer's fingertips and are centered on the individual. Mobile network technology has accelerated the flow of information and coverage in ways never seen before. Data exchange is freed from any geographic limits, allowing the maximum variation in user experiences. The Chinese firm Xiaomi leverages the social media platform Weibo to interact with users, making a user-oriented focus the driver of its business strategy, allowing customers to participate and voice their opinions, and eventually buy their own ideas. This user-oriented business model will be a part of the mainstream for a long time to come. It will generate a business pattern that is segmented, personalized, engaged, experiential, and fast. Consumers will have a vast range of products available to purchase instantly. Everything will be only a click away. Consumers will start to redefine the value of things in their own ways.

As I mentioned earlier, this seemingly personalized consumer consumption is essentially limited. Why is that? The core reason is that once consumer behavior is digitized, consumer preferences are datafied, both our information production and reception become data-driven, and traffic becomes the core of information. This means that Internet commercial platforms control and dominate user behavior data and the corresponding authority to push information. Thus, what appears to be a personalized consumption model will ultimately lead us into an era of limited personalization, controlled and driven by algorithms.

## More Focus on Product Design

People are no longer satisfied with traditional values that focus solely on the functionality of products. The design and image of products have become a strong factor driving value in meeting modern consumer needs. Emotional value, a sense of attachment, is something many people have begun to look for in products as well. This means that once the functional utility of a product is satisfied, what can truly move consumers are the emotional and cultural aspects, the visual and sensory stimulation, and the emotional resonance they bring.

For any commercial activity, whether it's flat advertisements, posters, digital films, or actual products, design is the most direct form of expressing commercial intent. For tangible products, the design of the product's appearance is the most direct way to persuade consumers. In an era of material surplus and when consumers have lower desires, design will become one of the core competitive strengths in business competition. The spirit and connotation of a product can only be effectively communicated through design, and consumers recognize whether a product meets their consumption needs through its design. In other words, many people have moved from basic material consumption to consumption at a spiritual level, namely design consumption, which is particularly prominent in the consumption of luxury goods.

Furthermore, consumers' personalized demands will increasingly be met through purchasing design services. Even the most attractive product will lose its uniqueness and, consequently, its value if everyone has it. This is an era that values uniqueness and individuality, where "my space, my rules" prevails. If your product is unique, especially if it's tailor-made for me, I would be very willing to lower my requirements for its price and practicality. Especially as 3D printing technology matures and personalized customization becomes more convenient and straightforward in the new era of industrial production, on-demand personalized design and customization will become the mainstream method in new business models.

## Moving from a Singular Shopping Pattern to a Comprehensive One

People once moved from place to place in pursuit of one specific commodity that was needed. Now it is more common to shop in one place for all that is needed. Big shopping malls have become part of modern life. Online, at a click of a button, we not only have the world's commodities in front of us, but we can also have them delivered to our doorstep thanks to dramatically improved distribution networks, or even unmanned delivery 24 hours a day. However, this is not yet at the level that can be called "comprehensive shopping." True comprehensive shopping means that a consumer completes a complex shopping process within a very short period of time.

A comprehensive shopping experience allows the integration of the purchasing of a range of commodities—from virtual products to physical, online to off-line—cutting across needs at home or at work, all based on the use of mobile Internet devices to enable the purchase of an extensive selection of products anywhere and at any time. One early example of a comprehensive shopping experience is provided by Alipay from China. Alipay has utilized its own platform to enable users to register for medical consultations, monitor queues, conduct payments, review results, and more. Alipay will establish a comprehensive online platform to allow virtual mobile prescriptions, medicine deliveries, hospital transfers, and medical insurance reimbursements, as well as commercial insurance and damage claims. It will then further utilize its big data platform and cloud computing capability to work with wearable technology manufacturers, medical care institutions, and even government agencies to construct a healthcare management platform. It will facilitate a shift in healthcare from treatment to prevention in the future.

The future may be one with one-stop shops that deal with all our needs. This may even be the case for medical care, eliminating queues and reducing waiting times. Indeed, when combined with AI generative large models, when we plan to travel, all we need to do is inform the AI of our travel dates, plans, and budget requirements. It can then provide us with an optimal travel itinerary, including the best times to travel, the most suitable flights, the most appropriate travel routes, and hotel options that meet our criteria. Once we approve the plan, the AI can assist us with a series of bookings and payments, streamlining the entire process and making travel planning more efficient and tailored to our preferences.

PART ONE

# WEARABLES, FROM THE FUTURE TO THE PRESENT

Smart wearables are essentially wearable smart sensors, spanning from humans to objects, and are a core carrier for constituting the metaverse. Wearable devices typically refer to intelligent monitoring devices that can be worn on the human body, such as smartwatches, smart clothing, smart glasses, and smart shoes. These devices possess various functions found in current smartphones, tablets, and PCs, but their most significant difference lies in two aspects. On the one hand, in terms of form, wearable devices primarily adopt product forms that conform to comfortable human wear. On the other hand, wearable devices are embedded with various high-precision and sensitive sensors, serving as input terminals that achieve an unprecedented level of integration with the human body. When wearable devices become sufficiently mature, they will become a part of our lives and even our bodies. For instance, they can be glasses, bracelets, watches, apparel, socks, underwear, hats, or anything closely related to our daily lives.

In the future, every part of the human body could become a potential field for the development of wearable devices. Beyond the obvious areas like the head, wrists, and feet, numerous micro-sensing devices implanted inside the human body are breaking new ground, bringing another round of significant transformations to social production and life.

CHAPTER 1

# UNDERSTANDING WEARABLES

Wearable devices represent a technology concept that has transitioned from science fiction to reality, from the future to the present.

We have seen the presence of wearable devices in many science fiction works. For example, in the 1979 film *Iron Man Breaks through Space City*, the watch worn on the protagonist's wrist could not only converse but also transform into a watch bomb. Ten years ago, in *007: Skyfall*, James Bond used a smartwatch capable of waterproofing, filming, and recording to "cheat" and successfully uncover a series of secrets. In *Mission: Impossible—Ghost Protocol*, Tom Cruise's high-tech smart contact lenses, and the communicators in *Star Trek* are other examples. These devices, wearable directly on the body or integrated into our clothing or accessories, represent what we call wearable devices.

## 1.1 THE BIRTH AND DEVELOPMENT OF WEARABLES

Surprisingly, the birthplace of the earliest wearable devices was the casino.

Massachusetts Institute of Technology mathematics professor Edward O. Thorp mentioned in his gambling guidebook, *Beat the Dealer*, that in 1955, he conceived an idea about a wearable computer to improve the odds of winning at roulette. Based on this idea, in 1961, Edward O. Thorp and another developer, Claude Shannon, collaborated to create this device. It was a small wearable computer that, when worn, could calculate probabilities during gambling to verify their formulas. With this wearable device, Edward O. Thorp successfully increased his roulette winning odds by 44%.

Afterward, the wearable device field began to develop rapidly, with multiple categories emerging during this period. The world's first wrist calculator, Pulsar, was officially released at the end of 1975, sparking a trend. It was rumored that US President Ford showed interest in the limited edition Pulsar priced at USD 3,950, which excited the media, although Ford later stated it was just a family joke.

In 1981, Steve Mann, still a high school student and later known as the "father of wearable technology," designed the first-ever head-mounted camera, connecting a computer to a backpack frame to control photographic equipment. The display screen was a camera viewfinder attached to the helmet.

In 1984, Casio introduced the world's first digital watch capable of storing information, the Casio Databank CD-40. In 1989, Reflection Technology launched the Private Eye head-mounted display. In 1994, researchers at the University of Toronto developed a wrist computer that could attach a keyboard and display screen to the forearm.

However, limited by technology, cost, application scenarios, and supporting facilities, these innovative products did not reach the consumer market or benefit the general public.

Take the Private Eye display as an example: resembling a head-mounted headset, it featured a 1.25-inch 720 * 280-pixel monochrome screen that could emulate the viewing experience of a 15-inch monitor. Apart from reading text documents, this wearable device had limited functionality and portability. The same was true for smartwatches. In 2004, Microsoft introduced the first smartwatch, MSN Direct, ushering in the era of the second screen. However, it only provided information reception functions like news, weather, and text messages—and even required users to pay a monthly or annual fee—without the ability to connect to a phone for notifications and messages.

Although these early wearable devices were quickly phased out, they formed the embryonic stage of the wearable field, and their avant-garde design concepts influenced subsequent products.

## 1.2 GOOGLE GLASS: IGNITING THE WEARABLE REVOLUTION

In the 21st century, wearable devices have entered a rapid development phase. In 2006, Nike and Apple jointly launched Nike+iPod—a sports kit that allowed users to sync their activities with their iPod. Nike also introduced several pieces of apparel with special pockets for the iPod.

In 2007, James Park and Eric Friedman founded Fitbit. In 2009, Fitbit released its first product—the Fitbit Tracker. In less than three years, this thumb-sized gadget sparked a fitness and health craze in North America, making personal portable wearable health devices a new favorite among venture capitalists.

In 2012, the Pebble smartwatch raised USD 10 million on the crowdfunding platform Kickstarter, far exceeding its original goal of USD 100,000. It attracted a huge following from tech enthusiasts and sports fans. This smartwatch, compatible with both iPhone and Android phone systems, allowed users to check iMessages from iOS devices directly on the Pebble watch. Furthermore, it could display incoming call information, browse web pages, and provide real-time alerts for emails, texts, tweets, and social network updates. Its simple, stylish design and variety of color options won the favor of consumers.

While these wearable products made some ripples in the wearable device market, the release of Google Glass is often seen as a milestone that had a significant impact on the development of wearable technology.

In 2012, Google shook the entire tech world, and even the world of fashion, with its revolutionary Google Glass. Sergey Brin, co-founder of Google, wore the device as part of New York Fashion Week. This device, resembling a pair of glasses, is equipped with a camera and a strip-like computer processor on one side, enabling functionalities similar to those of a smartphone. Users can control it via voice to take photos/videos, make calls, navigate, browse the Internet, and manage texts and emails. Suddenly, dazzling advertisements for Google Glass were everywhere, and for a time, Google Glass was considered a product that represented the "future."

On October 30, 2013, images of the second generation Google Glass were released on Google+. The first generation adopted bone conduction transducer technology to transmit sound. The second generation added earphones.

On April 10, 2014, Google announced the online sale of Google Glass, which is scheduled for April 15. Any US resident over 18 is eligible to purchase one at the price of USD 1,500. The sales window was only one day.

On May 25, 2014, Google started to sell the Explorer Edition of Google Glass to all Americans over 18 years of age. It could be purchased directly from the official Google website.

On June 23, 2014, Google announced the first international launch of Google Glass, which is now open to residents of the United Kingdom. Its selling price in the UK would be GBP 1,000 (USD 1,713 at the time).

In July 2014, Google Glass officially launched its live-streaming feature, showcasing Google's leadership in technological innovation. Google wasn't just ahead in the realm of smart glasses but also one of the first companies to explore the live-streaming business model. When the live-streaming feature was activated, Google started offering the livestream video-sharing app in its MyGlass store. Wearers of Google Glass equipped with this app could simply say, "OK, Google Glass, start broadcasting," to share their experiences with others on Livestream for free, whereas the app had been in the testing phase prior to this.

The software released for Google smart glasses could serve as a surgical teaching tool in medical schools. Surgeons could wear Google Glass to live-stream their operations, allowing students to watch the procedures through video without needing to be inside the operating room. Additionally, users could share their experiences at concerts or football matches through the device.

On November 25, 2014, Google decided to close its physical retail shop Basecamp, where Google Glass was sold. Most users were using the Internet or telephone for purchases and technical support.

That year, despite the limited production of Google Glass leading to inflated prices in the scalper market, there was still significant enthusiasm worldwide from early adopters. However, as the actual devices began to reach consumers, the reality of this overhyped product became apparent: excessive power consumption, discomfort when worn, features not significantly more impressive than those of a smartphone, and a lack of engaging applications. Particularly, the underdeveloped AR (augmented reality) application ecosystem for smart glasses was a major "stumbling block" preventing Google Glass from becoming a hit product.

A more significant issue was associated with the front-facing camera of Google Glass. Due to its design, the camera did not have any special indicators when taking photos or recording videos, which led to concerns about privacy. People couldn't help but wonder, "Is that Google Glass recording me?" This was particularly problematic in the West, where there was a strong emphasis on privacy and portrait rights. The fervent enthusiasm of some tech geeks exacerbated the issue. The infamous photo of tech blogger Robert Scoble wearing Google Glass in the shower indeed left many with psychological unease. Consequently, a new term, "Glasshole," emerged for Google Glass users, and several restaurants even declared they would not serve customers wearing Google Glass.

In a public opinion survey conducted by the market research firm Toluna, 72% of respondents cited privacy concerns as their reason for rejecting Google Glass. They were worried that hackers could access personal data through Google Glass, leading to the leakage of personal information, including location details.

In 2015, the Google Glass project was halted. Although it eventually made a comeback targeting industrial applications two years later, it's undeniable that Google Glass had gradually faded from the public eye. Only time will tell when or if it will make a significant return to the consumer market. Nonetheless, Google Glass, once a global leader in smart glasses, ignited the smart wearable industry, showcasing the advent of the smart wearables era and laying the groundwork for the concept of the metaverse.

## 1.3 THE SUCCESS AND FAILURE OF GOOGLE GLASS

From the moment Google Glass was born and entered the public eye as a practical application product, the fate of the entire wearable industry seemed to ebb and flow with Google Glass, from initial high expectations to subsequent criticisms and controversies. Regardless of how the media or consumers viewed it, Google Glass's position as the harbinger of a new era in wearable devices is undeniable; its inception was indeed grand.

Although Google Glass had a tumultuous journey in the consumer market—facing criticism for its appearance and collective resistance due to privacy concerns, even to the point of disappearance, leading many to view it as a failure—I see it differently.

In fact, Google Glass was always an experiment in the consumer market, plagued not just by aesthetic criticisms but also by significant privacy concerns. Before its launch, Google Glass had undergone numerous revisions and iterations within Google's labs, facing countless internal failures. From an initial idea about smart glasses to the first cumbersome prototypes that were impractical to wear, numerous updates and upgrades followed. Not every iteration was successful or represented a significant leap forward; often, they were just small steps of progress. However, Google persisted with a determination that exceeds that of many entrepreneurs, supporting the development of this "dream." Ultimately, Google Glass emerged as a wearable device.

The version of Google Glass we are familiar with is just one iteration of Google's ongoing development of the smart glasses project. At that point in time, Google's decision to announce and showcase this smart glasses product was based on its judgment of the overall technological development trend, foreseeing the imminent arrival of the metaverse era and the rise of the wearable industry. Thus, the proactive launch of Google Glass served not only to ignite the industry but also to conduct real-world commercial testing.

Initial testing started with the mass consumer market and applications in media, education, social networks, entertainment, and other fields. Then, it moved on to remote medical treatment, trials in the UK market, and enterprise-level applications. Google is still looking for an optimum commercial model for Google Glass.

Previously, Google X relied on internal experts to perfect the product concept and refine its design. To further capitalize on its commercial value, more work needs to be done in practical application scenarios. It is important to know the limits of Google Glass and its applications, as well as specific technical requirements for different needs and uses of the device.

This latest patent acquired for Google Glass is a newer version based on user tests. The previous Google Glass versions all used a spectacles-based format, which is not preferred by some users. This new format has made smart glasses intangible, allowing them to be worn by people who are not into wearing spectacles.

Google launched a Glass at Work program aimed at making apps for businesses to improve working conditions and productivity. In June 2014, Google announced the first five Glass at Work-certified partners. They were APX Labs, Augmedix, CrowdOptic, GuidiGO, and Wearable Intelligence:

(1) APX Labs—creator of Skylight business software for Glass, enabling instant access to business data at work
(2) Augmedix—developers of the app allowing doctors to access patients' medical data, including heart rate, blood pressure, and pulse rate, on Google Glass
(3) CrowdOptic—providers of content for live broadcasts and context-aware applications for sports, entertainment, building/security, and medical industries
(4) GuidiGO—partnering with museums and cultural institutions to help people connect with art and culture through story-telling and other experiences
(5) Wearable Intelligence—creators of Glassware for energy, manufacturing, healthcare, and more

As can be seen from the five certified partners of Glass at Work, their activities cut across many different areas and walks of life. Each of their apps has a clear purpose designed for precisely targeted professions. They require users to have certain professional knowledge. The global oilfield service giant Schlumberger has worked with Wearable Intelligence to equip its technicians with a modified version of Google Glass for quicker access to stock information and improved efficiency.

While facing setbacks in the consumer market, Google Glass embarked on a new journey in the ToB sector, finding its niche among numerous large manufacturing companies, albeit far removed from the public eye.

Even though Google Glass eventually receded from the market due to lackluster sales, Google was undeniably the pioneer in the realm of smart glasses. As a trailblazer in the wearable device industry, Google deeply understood that the application scenarios and commercial value of eyewear-based wearable devices would far exceed those of smartwatches and fitness bands. That is, in the wearable device industry, the currently popular smartwatches and fitness bands are actually the categories with the smallest application scenarios, while smart glasses and smart clothing are set to be the next major areas of growth in wearable technology.

In terms of product design, the design principles of Google Glass are still being adopted in the industry today, such as integrating a camera and GPS to enable outdoor AR experiences. Similarly, using the temple area of the glasses for touch control remains one of the mainstream interaction methods in the industry. The issue with Google was that its vision was too far ahead of its time, embodying the adage: "Being one step ahead is being a pioneer; being two steps ahead is being a martyr."

## 1.4 THE ULTIMATE AIM OF GOOGLE GLASS

When Google Glass Explorer was discontinued amid concerns about privacy, many thought Google was failing in the wearable tech industry. That is because many people do not entirely understand what these global tech giants are trying to achieve. Their understanding of Google Glass was wrong too.

The underlying reason Google entered the wearable tech industry was to control mobile Internet data access and to establish a mobile Internet-based big data search platform. For example, with the acquisition of Nest, Google is not trying to have a presence in the smart home industry but to likewise establish a big data search platform from it.

Building a big data platform is the aim of all these efforts in hardware development and acquisitions, as the big data platform is the foundation for a mobile Internet user base. Google, as we all know, is a search engine company, a big data stronghold in itself. Would Google leave its main business behind for a directional change to focus on a physical industry such as wearables or smart homes? Obviously not.

What drives Google to invest so much in developing Google Glass and its ongoing refinement and upgrades? We may need to look back at the beginning of this wearable revolution. Who kicked it off first? It was indeed Google through Google Glass. And who fueled the smart home concept? Again, it was Google, with their acquisition of Nest.

While others followed suit to build their businesses in this area, Google seemed to quietly move aside to focus on building a system platform instead. Google wants to be ready to present its dedicated operating systems for wearables and smart home app developers to use. Other players will all need such smart device app systems after conquering technical issues around smart home devices and wearables.

Hopefully, you can now see Google's intention much more clearly. It is not surprising that Google dismissed the entire Glass developing team after their mission was accomplished: to launch the wearable tech revolution. Google decided to develop its own Glass because it was in need of a device to support the construction of a search platform in the era of the mobile Internet. Google was able to perform repeated tests with its own products so that it could accumulate knowledge and refine the mobile Internet big data platform.

In the era of desktop personal computers, user retention was counted in hours or days. We could control the users only with the data platform of those PC terminals. It is a different thing with the mobile Internet. User retention is now counted in minutes. It is easier now to go about without a PC, but much harder to go without a mobile phone. We now can see a major difference between mobile Internet and wired Internet services—the sticking time is much shorter.

This will be further shortened with the emergence of wearable devices. To safeguard its big data platform stronghold, Google must focus on user stickiness. It is very clear to Google that ultimate user stickiness in the era of the mobile Internet is achieved through coding the operating system of wearable devices.

Google Glass was never the true aim. The commercialization of Glass also seemed halfhearted. Google is more interested in exploring the potential and guiding the direction of the wearable industry, including advances in medicine and healthcare.

Whether a commercially successful product or not, Glass fulfilled its mission to attract global interest, capital, and talent to the nascent business of wearable technology, among these new wearables, a significant part of them will use the system platform developed by Google. Their users will then build a big data empire through their wearable devices. The commercial success of Google Glass should not be measured by sales of the hardware device itself but rather by the advantage it has given Google in the exploitation of big data from mobile Internet users.

The seeming failure of Glass in the consumer market was, in fact, only an attempt to test a lab-developed product in a consumer context.

Another contribution brought by Google Glass is the debut of wearable-based VR and AR products.

Google Glass ignited the wearable market, and since its debut, an increasing number of wearable products have emerged in our lives, subtly transforming them. It was the advent of Google's smart glasses that paved the way for Apple's first head-mounted display, Apple Vision Pro, released on June 6, 2023. This new generation electronic product, equipped with multiple cameras, allows users to operate and control it using gestures, eye movements, or voice and can be used for work, entertainment, and communication. From Google's release of smart glasses in 2012 to Apple's launch of the Apple Vision Pro in 2023, Google was ahead by a full 11 years. It was Google's vision that allowed the world to foresee an upcoming era of smart wearables, which subsequently influenced Mate's proposal of the metaverse concept.

Today, wearable devices have evolved from conceptual ideas to becoming a pivotal force in transforming various domains. These changes have unfolded over just 50 years, a period marked by debates and stagnation, but wearables have inevitably arrived. Their unstoppable momentum is not solely because they encapsulate high-tech advancements but because they endow the Internet with physical attributes, becoming a new portal for data flow in the mobile Internet era and making the realization of the metaverse concept possible. In the foreseeable future, wearable devices and the underlying software, applications, and content services will become indispensable parts of everyone's daily life and work, signifying a trend and current in the future development of human society.

CHAPTER 2

# WEARABLE PRODUCTS

In recent years, the concept of wearable devices has remained popular, with a plethora of wearable products emerging. Currently, the market features a variety of wearable product forms, including smartwatches, fitness bands, smart glasses, and smart clothing. In terms of product functions, medical health, information entertainment, and sports health are hot spots and trends.

## 2.1 SMARTWATCHES

Among all wearable devices, smartwatches are undoubtedly the most familiar to the public and the most commercially mature product type.

In April 2012, the wearable startup company Pebble launched a crowdfunding campaign on Kickstarter for its smartwatch, Pebble. In an environment where there was little concept of smartwatches, Pebble, with its e-paper screen design that balanced battery life, captured the attention of many users. By the end of the crowdfunding campaign, Pebble had raised over USD 10 million.

In the following years, after fulfilling its first batch of crowdfunding orders, Pebble prepared for a second crowdfunding campaign and a series of new products, releasing Time Steel, Time Round, and other models. The e-paper screen was upgraded from black and white to color, and in 2016, Pebble started to focus on fitness features.

Regrettably, like some other crowdfunding stars, Pebble experienced its moment of glory shortly after starting, only to later encounter the production and distribution challenges that many startups face.

The real breakthrough that brought smartwatches to the market and ignited it was in 2015 with the birth of the Apple Watch, which then received annual updates. Although Apple had already released some information about the Apple Watch in 2014, the announcement still dominated headlines across major media outlets.

Apple's first smartwatch came in three styles: the standard Apple Watch, the Apple Watch Sport, and the luxury Apple Watch Edition. It supported making calls, responding to texts with voice, connecting to cars, obtaining weather and flight information, offering map navigation, playing music, and measuring heartbeat and steps, among dozens of other functions. It could rightly be described as a comprehensive health and sports tracking device.

In addition to launching the Apple Watch, Apple introduced a new mobile application platform at its developers' conference, designed to collect and analyze users' health data, which they named "HealthKit." Apple executives explained to developers that it could integrate data collected by various health apps on iPhone, iPad, and Apple Watch, such as blood pressure and weight.

Following HealthKit in 2015, Apple created a software infrastructure specifically for medical researchers aimed at addressing current challenges in medical surveys, such as a lack of sufficient samples and participants and inadequate data support. The platform's greatest value lies in assisting researchers and medical professionals in collecting and organizing patients' medical data through the iPhone, aiding in the diagnosis of various diseases. Current applications already encompass breast cancer, diabetes, Parkinson's disease, cardiovascular diseases, and asthma. Users can also monitor and track their own physiological data using the Apple Watch.

At the launch event, Tim Cook officially stated that using the Apple Watch could lead to a healthier lifestyle. The Apple Watch's Activity app displays users' daily physical activities, including calories burned from walking and exercise time, providing timely feedback to users. Additionally, users can set appropriate goals, and the Apple Watch will remind them daily and assist them in achieving those objectives.

After its release in 2015, the Apple Watch spread rapidly, reaching a shipment volume of 11.6 million units in just nine months, whereas the total annual shipment volume of smartwatches in 2014 was less than seven million. In the following years, the Apple Watch made significant strides, even surpassing Rolex, a giant in the traditional watch industry, to become the highest-grossing watch globally in 2017.

In 2022, Apple remained the largest smartwatch brand worldwide. According to Counterpoint Research, Apple held a 34.1% share of the global smartwatch shipment volume in 2022, achieving a 1.5% year-over-year growth. Counterpoint attributes Apple's growth mainly to the Apple Watch Series 8, Apple Watch Ultra, and Apple Watch SE 2022 edition. The Apple Watch Ultra, targeting extreme and outdoor sports enthusiasts, expanded Apple's consumer base in the smartwatch sector. As the "king" of the smartwatch market, Apple's shipment volume grew by 17% year-over-year, accounting for 60% of the global smartwatch market revenue in 2022.

The success of the Apple Watch opened up the market for smartwatches, prompting various manufacturers to enter and strategize in the smartwatch market, such as Huawei and Samsung.

Starting with the Watch GT in 2018, Huawei abandoned Google's Wear OS in favor of its LiteOS system. With its proprietary chip architecture, Huawei can integrate hardware and software deeply, enabling longer battery life and continuous heart rate monitoring. In the same year, Samsung's Galaxy Watch Active came pre-installed with Samsung's in-house Tizen OS, which, thanks to system-level integration, allows the watch to quickly pair with Samsung smartphones via the Galaxy Wearable app on the phone.

In addition to smartwatches for adults, children's smartwatches have also garnered market attention. In mainland China, brands like 360 and Xiao Tiancai have become enduring memories in the era of children's smartwatches. As a pioneer in the children's watch market, 360 released its first product in 2014, focusing on creating children's watches with calling features as stand-alone devices.

Though Xiao Tiancai wasn't the first mover in China's mainland children's smartwatch field, its strong distribution capabilities quickly secured a significant market share, dominating the shipment volume for a long time. The popularization of smartwatches has also made the public more aware of and interested in the concept of smart wearables, leading to more attention and contemplation on wearable devices.

## 2.2 FITNESS BANDS

In addition to smartwatches, another popular wearable device for the wrist is the fitness band. One of the earliest players in this field was Jawbone. Initially, Jawbone focused not on wearables but on Bluetooth speakers and headsets, entering the wearable market in 2011.

In the wearable sector, Jawbone had only one product line: the UP series of fitness bands. The Jawbone UP was a wristband device that tracked daily activities, sleep patterns, and dietary habits, featuring smart alarms, idle alerts, special reminders, and nap modes. Its health tracking functionalities included various bioimpedance sensors, measuring heart rate, respiratory rate, galvanic skin response, skin temperature, and ambient temperature. Additionally, based on user input such as weight, height, age, and gender, the Jawbone UP could more accurately display steps, distance, speed, and calories burned during physical activities.

However, due to the lack of a screen and limited interactivity, fitness bands suffered from low user engagement. Unfortunately, Jawbone did not adapt or transform in time. High pricing put the company at a disadvantage, especially after Xiaomi launched its smart band, which was priced at RMB 79. Furthermore, Jawbone neglected its original Bluetooth speaker business due to the capital's focus on the wearable market. Eventually, in 2016, Jawbone exited the wireless speaker market and ceased production of its fitness bands.

Besides Jawbone, another early participant in the fitness band market was Misfit, which crafted fashionable wearable devices. Misfit released products like Misfit Shine and Misfit Ray, and in 2015, the well-known American fashion brand Fossil acquired Misfit for USD 260 million.

In addition, Fitbit, now acquired by Google, stands out as a venerable powerhouse in the fitness band domain. Unlike other wearable manufacturers with fluctuating fortunes, Fitbit was the reigning champion of its time. Despite facing pressure from giants like Apple and Google, Fitbit carved out its niche by focusing on specialized health features. In 2013, Fitbit launched the affordable and precise Fitbit Flex series. By 2014, Fitbit's fitness band shipments had reached 19 million units. In June 2015, Fitbit went public on the NYSE, becoming the first publicly traded company in the wearable sector. Over the following two years, Fitbit acquired several companies and projects, including Pebble, Fitstar (a fitness app), Coin (mobile payment), and Vector Watch (a British smartwatch company).

In China, Xiaomi quickly dominated the domestic market with its high-value-for-money fitness band series. By 2015, just a year after entering the wearable market, Xiaomi seized the top spot in China. Notably, Xiaomi's first fitness band was priced at just RMB 79, effectively becoming the "price butcher" of the fitness band market, overshadowing nearly all other products priced above RMB 100. The industry was even fearful of Xiaomi's low-price strategy.

For instance, the Chinese fitness band brand "bong" released its "bong2" band at RMB 99 the day after Xiaomi's launch, a steep drop from the first generation's price of RMB 699. Despite the founder's hopes to confront Xiaomi with a "David versus Goliath" spirit and various product experiments, the "bong" eventually exited the market. Similarly, the Betwine team, which launched a social fitness band, also faded away.

In terms of functionality, both smartwatches and fitness bands share several features with smartphones, such as time display, voice chat, mobile payments, text messaging, and information queries. However, these functions don't represent their true value. As the consumer electronics devices closest to our bodies, the future direction of smartwatches and fitness bands lies in further meeting the healthcare needs of users.

Today, especially with their unique advantages in health monitoring and the maturation of various biosensing technologies, smartwatches, and fitness bands, they can provide an increasing amount of detailed health data. Moreover, collaborations with medical institutions to develop more professional health monitoring features have become a common strategy among manufacturers of smartwatches and fitness bands.

## 2.3 SMART EARBUDS

In the wearable market, smart earbuds, especially True Wireless Stereo (TWS) earphones, represent a significant segment. It's becoming increasingly common to see people around us wearing wireless earbuds, making them one of the fastest-growing categories in wearable devices.

The first wireless earbuds appeared in 2014 when the world's inaugural truly wireless earbuds, Dash, debuted via crowdfunding. Developed by the German manufacturer Bragi, Dash aimed to become a "miniature computer for the ear," incorporating nearly all the popular technologies associated with true wireless earbuds. However, Bragi couldn't keep up with competitors like Apple and Jabra and announced its exit from the hardware market in April 2019.

By 2017, major manufacturers, both domestically and internationally, had joined the trend, creating their own TWS earbuds and headsets with integrated voice assistants. By 2019, wireless earbuds had become a mature digital product. They offer numerous advantages and a brighter future compared to wired earphones. As our lives continue to evolve with technology, and as online work, learning, entertainment, and daily life become the norm, human interaction with digital devices is only set to increase. Wearable devices like earbuds, no longer occasional accessories, have become extensions of our senses, playing a crucial role in how we capture information and engage with our daily tasks and leisure, assuming vital functions.

Of course, the most popular wireless earbuds are Apple's AirPods series, which significantly contributed to the popularity of wireless earbuds. In 2016, amid the stagnant smartwatch market, Apple launched its first truly wireless AirPods. Their unique design, combined with a seamless experience with the iPhone, quickly made AirPods a leading product in the wearable industry. The second generation of AirPods was introduced in 2019, featuring new functionalities like voice-activated Siri and wireless charging, signaling new industry trends.

Today, the AirPods lineup has evolved into the AirPods Pro second generation with active noise cancellation (ANC). Besides basic functionalities, its defining feature is ANC, which works by capturing ambient noise with microphones and generating an anti-noise signal to reduce the noise level. Human ears are particularly sensitive to noise in the 20–500 Hz frequency range, which is challenging to block with physical earbud barriers alone, making this frequency range the focal point of ANC technology competition.

In addition to Apple's AirPods series, other smartphone manufacturers like Huawei, Samsung, and Xiaomi, as well as established headphone brands such as Beats and EDIFIER, have also developed mature products in the smart earbuds category.

Moreover, the integration of large AI models has added numerous functionalities to smart earbuds, including intelligent assistants, fitness tracking, and real-time language translation. For instance, iFLYBUDS Nano+ from iFLYTEK incorporates a generative AI assistant called VIAIM.

While in-ear wireless earbuds continue to evolve, a new type of smart earbuds known as Open Wearable Sound (OWS) devices is rapidly emerging. The key difference between OWS and TWS lies in the design of the sound-emitting unit. OWS devices feature an open structure, meaning they do not penetrate deeply into the user's ear canal during use. Currently, OWS products mainly come in ear-hook and ear-clip styles, with some appearing as audio glasses.

Ear-hook OWS devices hang on the upper outside of the ear with their curved ear-hook structure, offering stability and more space to accommodate various components. This design is popular among mainstream OWS earbuds, such as Aftershokz's bone-conduction headphones.

Ear-clip OWS devices are more compact, with supports usually made of soft and elastic materials. This design is user-friendly for those who wear glasses, as it prevents conflicts between the product and the arms of the glasses. A notable example is Huawei's newly launched FreeClip.

Additionally, audio glasses can be considered a branch of OWS. Notable products in this category include Xiaomi's audio glasses, Bose Frames, and Huawei Eyewear II. These devices merge the functionality of smart earbuds with eyewear, offering users a unique way to integrate audio technology into their daily lives.

## 2.4 SMART GLASSES

When discussing wearable devices, smart glasses are an unavoidable topic. To a large extent, the wearable market was ignited by Google's smart glasses. Although Google Glass eventually exited the market, the impact and shock it brought to the wearable and smart glasses market are undeniable.

The current smart glasses market primarily combines smart audio with the form of glasses, equipped with independent operating systems to merge the functionalities of both. This allows them not only to serve as a fashion accessory but also to provide an open audio experience. However, this is just the beginning for manufacturers in the smart glasses arena. As technology evolves, a more significant trend for smart glasses will be their integration with vision to implement AR functionalities, offering richer applications.

### 2.4.1 Smart Audio Glasses

While smart audio glasses are just the beginning for smart glasses, many manufacturers are already positioning themselves in the market to gain a competitive edge in the future landscape of smart glasses. Brands like Bose, Amazon, Huawei, and Aftershokz have launched iterative products, while Rapoo, Razer, and others have introduced their first smart glasses products.

For example, the Amazon Echo Frames 2nd Gen can wirelessly connect to the wearer's smartphone, offering improved battery life compared to its predecessor, with four hours of continuous playback time. It supports an auto-off function by flipping the frames upside down for three seconds, and an echo frame feature that automatically adjusts the volume based on the ambient noise level. The 2nd Gen also has a VIP filter for customizing notification preferences. Amazon has enhanced the audio quality, promising richer and fuller sound for music and Alexa responses. Users can make calls, set reminders, add to-do lists, get news, listen to podcasts, or control their smart home just by asking Alexa. Besides the official Alexa voice assistant, it supports native voice assistants on iOS and Android smartphones. For privacy protection, users can completely disable the microphone by double-tapping the action button.

Bose Frames, Bose's smart audio glasses, look similar to regular sunglasses but feature a more trendy and minimalist design. The lenses are interchangeable, and the hinges are gold-plated, giving the glasses a high-end, fashionable appearance. Bose Frames utilize Bose's advanced directional audio technology, which allows the open-ear speakers to deliver sound directly to the wearer without others overhearing, enhancing the listening experience. Internally, the Bose Frames are equipped with the high-end Qualcomm CSR8675 Bluetooth audio SoC, supporting 24-bit audio transmission and the aptX HD audio format. This technology ensures superior audio quality for the Bose smart audio glasses, making them not only a stylish accessory but also a high-performance audio device.

The Huawei X Gentle Monster Eyewear II smart glasses haven't undergone significant changes in appearance compared to their previous generation. The main change is in the glasses case, which has shifted from a pouch to a box style, and the integrated battery in the charging case has been removed. In response, NFC wireless fast charging technology has been adopted to reduce charging time and enhance the capability for quick power replenishment. The battery life of the glasses has been significantly improved, with overall endurance doubling, allowing for five hours of continuous use.

Other major changes include the main control chip and speaker unit. The Huawei Eyewear II smart glasses have replaced the previous generation's BES Hengxuan 2300 Bluetooth audio SoC with Huawei's self-developed HiSilicon Hi1132 main control chip, and the Bluetooth version has been upgraded to

5.2. For the speaker, the Eyewear II uses a 128 $mm^2$ custom diaphragm and is supported by an inverse sound field acoustic system, which effectively reduces sound leakage and enhances the wearing experience.

### 2.4.2 AR Smart Glasses

Beyond smart audio glasses, brands like Apple, Facebook, and OPPO have initiated projects in AR smart glasses.

Many are familiar with VR technology, which offers a fully immersive experience where users interact with a virtual environment. However, VR's immersive nature limits its mobility—users must ensure they are in a safe environment to prevent collisions or falls. In contrast, AR technology overlays digital information onto the real-world, offering a seamless integration of real-world and virtual information. AR's real value to users is most evident in mobile scenarios. For instance, in unfamiliar settings, AR can provide additional information about the surrounding environment or guide users to their destinations, making it an ideal complement to mobile networks. Smart audio glasses represent a preliminary stage in AR glasses development, initially addressing open spatial audio. The transition to commercial, portable AR smart glasses will revolutionize the user experience with electronic tech products.

Google Glass is an example of AR glasses, employing an optical see through (OST) method. It allows the physical world to be viewed through optical perspectives, overlaying digital information via the device's optics, thus achieving AR. AR's essence is the fusion of virtual and real, and OST's high-transparency optical solution allows virtual images to be projected directly into the real-world, reducing device latency, positioning requirements, sensor demands, computational power, and energy consumption. OST emerges as the optimal solution for AR compared to cumbersome head-mounted displays.

Besides Google Glass, subsequent devices like Microsoft's HoloLens and the startup Magic Leap have also adopted this approach. Microsoft launched HoloLens in 2015, offering high-definition holographic images, spatial sound, and voice/gesture control for a refreshing AR experience. Despite facing potential discontinuation, Microsoft decided to focus HoloLens on industry applications, updating it to the second generation in February 2019. To date, the HoloLens series remains a benchmark in the AR glasses field.

Additionally, Thunderbird's Thunderbird X2 and X2 Lite are AR glasses based on the OST approach. Released in 2023 by Thunderbird Innovation, Thunderbird X2 is the world's first mass-produced and commercially available binocular full-color MicroLED waveguide AR glasses. It boasts over 85% lens transparency, a peak brightness of 1,500 nits, and a 3D full-color display, ensuring clear visibility even outdoors. The Thunderbird X2 Lite, an iteration of X2, has undergone significant weight reduction. The full-color light engine is dramatically downsized, reducing the weight by 30%–40%. Combined with breakthroughs in waveguide design, material innovation, and new industrial design, the X2 Lite's total weight has dropped from 119 g to about 60 g. This weight reduction places it below many fashionable plate glasses, meaning it can be worn all day without burden.

Meanwhile, many brands like Apple, Huawei, Samsung, Microsoft, Facebook, and OPPO are actively exploring AR smart glasses. However, compared to smart audio glasses, due to technological challenges and limitations in size and weight, many of these AR glasses are still in the development and conceptual stages and are not yet as portable and wearable as regular glasses.

Among these, Apple's AR glasses have garnered significant market attention. As early as 2018, reports predicted that Apple would release its AR glasses, dubbed "Apple Glasses," by 2021 at the latest. Apple has applied for dozens of software and hardware patents in AR device development, and many of Apple's technologies and features are seen as groundwork for AR devices, such as AirPods' spatial audio technology, UWB ultra-wideband chips in iPhone / Apple Watch / HomePod, and the LiDAR scanner in the iPhone series and iPad Pro.

In 2023, Apple introduced a mixed reality (MR) headset device—Vision Pro. According to the introduction at the launch event, this "spatial computing" headset, equipped with a 4K display, allows users to freely switch between VR and AR using a dial. Besides Apple's previous generation M2 chip, Vision Pro also includes a custom-designed R1 processor and 16 GB of unified memory. Supported by the new visionOS operating system, users can control the device with just their eyes, hands, or voice. Vision Pro can project a Mac screen into the air, serving as a portable display for multitasking. Furthermore, apps, photos, and videos from iPhone and iPad can also be freely viewed.

As Apple reveals more columns and products, its AR glasses seem imminent, likely to bring another groundbreaking disruption to the smart glasses market, similar to the impact of the Apple Watch's release years ago.

Facebook also announced in 2020 that it was developing VR glasses, which differ significantly in appearance from the current "bulky" VR headsets, resembling conventional sunglasses. These glasses, less than 9 mm thick, utilize holographic optics and other technologies, though the project is said to be in a purely research phase.

Samsung has been continuously exploring AR smart glasses as well. Previously leaked promotional videos for Samsung's AR glasses showed the functionalities of the Samsung Glasses Lite, including watching movies on a projected screen, gaming, office work, and connecting with phones and drones to serve as a display. Another video showcased future applications for Samsung's AR glasses, merging reality and virtual imagery to enable VR scenes and virtual video calls.

## 2.5 SMART HEADSET

Oculus initiated a Kickstarter campaign on 1st August 2012 to fund Oculus Rift, a VR headset. The campaign's tagline promised that Oculus Rift "will change the way you think about gaming forever." Since then, it has amazed gamers with its performance at this stage.

The Oculus Rift is a VR headset designed specifically for video games. Its unique performance immediately won the hearts of many. In less than a month, its crowdsourcing campaign racked up USD 2.43 million from more than 9,000 backers looking to fund its development and manufacture.

The two landscape displays in Rift provided a combined 1280 × 800 resolution, with each eye seeing an image in 640 × 800. The biggest feature of the Rift is the gyroscope-enabled viewing field, which significantly improves the sense of immersion. Oculus Rift headsets could be linked to computers or gaming consoles via DVI, HDMI, or micro USB.

On March 26, 2014, Facebook announced its acquisition of Oculus VR for around USD 2 billion, taking an important step into the wearable tech industry.

Today, the Oculus brand has undoubtedly become a titan in the VR headset market. According to data released by the Steam platform, as of March 2021, the top four brands on SteamVR were Oculus, HTC, Valve, and Microsoft WMR. Oculus dominated with a staggering 58.07% market share. Following the launch of Oculus Quest 2, its market share surged, crowning it as the top VR headset on Steam in February 2021. In March, its dominance continued, with its share expanding to 24.25%, maintaining the top spot on SteamVR's most active VR devices for two consecutive months. As Facebook's latest generation all-in-one VR machine, Oculus Quest 2 had an impressive debut in September 2020, with pre-orders reaching five times that of the original model. According to Andrew Bosworth, Vice President of Facebook Reality Labs, within less than six months of its release, cumulative sales had already surpassed the total of all previous Oculus VR headsets combined.

Similar to Oculus' VR headsets is the HTC VIVE series, developed jointly by HTC and Valve. The first developer version of VIVE was unveiled at MWC in 2015, with the consumer version hitting the market in 2016. According to Steamspy data, sales of this product neared 100,000 units within three months of its release.

Additionally, in September 2022, ByteDance released its first VR headset, PICO 4, marking its flagship VR product since acquiring PICO. There is anticipation for PICO 4 to deliver an enhanced VR experience, propel the VR industry forward, and penetrate the global market. At the new product launch, Zhou Hongwei, the founder of PICO, introduced two consumer-grade products: PICO 4 and PICO 4 Pro. Both headsets are powered by the Snapdragon XR 2 platform and employ a foldable Pancake optical path, supporting intelligent IPD adjustment and hand tracking. The primary distinction between the two is that PICO 4 Pro adds three near-infrared cameras inside the headset, enabling automatic IPD adjustment, eye tracking, and facial tracking.

Sony and Samsung have also ventured into the smart headset arena. However, current smart headsets are predominantly developed for gaming. As the technology evolves, VR's unique advantages in other sectors are becoming evident, particularly in healthcare. VR technology can create a virtual natural environment using computers and specialized software, aiding doctors in diagnosing diseases and providing realistic training. For example, during training, doctors can interact with a simulated patient, which reacts based on the doctor's correct or incorrect actions.

VR headsets allow us to virtually experience places unreachable in reality, offering experiences beyond the constraints of space, time, and medium. Wearing a VR headset marks the beginning of our immersion into another world, even becoming a part of that world. The future of VR technology extends well beyond gaming, poised to penetrate various sectors, especially healthcare and travel, where its impact will deepen significantly.

## 2.6 SMART RINGS

Smart rings are emerging as a new favorite in the consumer electronics market, integrating health wearables and spatial interaction. As major wearable players make their moves and companies like Apple and Samsung gear up, alongside XR manufacturers strategizing their entries, smart rings are increasingly capturing public attention. Notable products in this category include the Oura Ring, Ultrahuman Ring AIR, and Luna Ring.

### 2.6.1 Oura Ring

Oura was established in 2013 and successfully completed its crowdfunding on Kickstarter, the largest crowdfunding site in the United States. The first generation of Oura Ring was released in 2015, and to date, there have been three generations of Oura Rings. Oura Ring aims to help users unlock their potential, foster health awareness, and promote healthier lifestyles. The original trio of co-founders comprises backgrounds in engineering, design, and data science.

In 2020, a story where an Oura Ring fortuitously warned a user of potential

COVID-19 symptoms caught the attention of the NBA, leading to an official partnership. The NBA purchased 2,000 Oura Rings for staff and players participating in the season restart. Oura provided NBA players with COVID-19 infection alerts and other health metrics monitoring, aiding in improving their sleep quality and readiness for games. Before this collaboration, Oura was mainly known within Silicon Valley elite circles, but the partnership with the NBA propelled it into the sports world and the public eye. That same year, Oura established partnerships with the WNBA (Women's National Basketball Association), UFC (Ultimate Fighting Championship), Formula 1's Red Bull Racing Team, and the MLB's (Major League Baseball) Seattle Mariners, enhancing its visibility and influence while refining its sensing and monitoring technology through serving athletes.

In 2022, Oura collaborated with the luxury brand Gucci to release a co-branded smart ring. Based on the third generation of Oura Ring and incorporating Gucci's design, this collaboration added a fashion attribute to the wearable device. Oura also entered Japanese telecom company SoftBank's off-line stores and e-commerce platforms, providing customers with tangible experiences of the online product. Additionally, celebrities like Gu Ailing and Prince William wore the ring, and a review video by the popular YouTuber Unbox Therapy garnered over a million views. With these various channels, Oura sold 500,000 rings in ten months, surpassing a cumulative shipment of one million rings.

The sensors embedded in the Oura Ring include a PPG sensor, temperature sensor, and a 3-axis accelerometer, focusing on sleep, heart rate, blood oxygen, body temperature, and activity metrics. Sleep monitoring is a flagship feature of Oura Ring, assessing different sleep stages (light, deep, REM) to enhance sleep quality and overall well-being, providing a superior nighttime wearing experience compared to smartwatches, and establishing itself as a veritable "sleep lab on your finger."

### 2.6.2 Ultrahuman Ring AIR

Ultrahuman was founded in 2019 by a group of biotechnology "hackers" who were inspired by observing how cutting-edge technology could enhance athletes' performance. The idea evolved to use algorithms and physiological data to monitor health and improve metabolism. The team, comprised of serial entrepreneurs committed to improving human health, has received support from top global venture capital firms like Alpha Wave and has established partnerships with manufacturers and logistics companies in five countries. Ultrahuman secured USD 7.6 million in Series A funding in December 2020 and USD 17.5 million in Series B funding in 2021. In April 2022, Ultrahuman acquired a wearable company, Lazyco, to expand into the biowearable field. In September 2022, the Ultrahuman Ring launched on Kickstarter, achieving tremendous success by raising over USD 500,000, surpassing its funding goal by 1,091%. They received 2,500 orders and have shipped over 90% of them.

The Ultrahuman Ring AIR, launched by the Indian health tech company Ultrahuman, emphasizes lightness, as suggested by the name "AIR." Efforts have been made to ensure comfort in wearability compared to the Oura Ring. The outer ring is made of pure titanium, coated with tungsten carbide to prevent scratches. The design is compact and seamless, with a super-thin, smooth inner band without protrusions, ensuring unmatched comfort throughout the day and night. The Ultrahuman Ring AIR comes in four colors: Astra Black, Matte Grey, Bionic Gold, and Space Silver, with additional sizes including 5 and 14, and a thickness of 2.45–2.8 mm. Due to material selection and structural design optimization, the ring weighs between 2.4–3.6 g.

The sensors embedded in the Ultrahuman Ring AIR are almost identical to those in the Oura Ring, with the difference being its PPG uses a single-channel sensor, and a six-axis accelerometer replaces the three-axis one, considering its focus on fitness applications.

Mohit Kumar, CEO and co-founder of Ultrahuman, which started with CGM (Continuous Glucose Monitoring), stated that if the ring wearer also uses Ultrahuman's CGM M1, the data collected by the ring's sensors would be linked to real-time glucose levels. This connection allows users to correlate glucose fluctuations with triggers like "high stress, poor sleep, low activity levels," enabling adjustable correlations. Compared to other smart rings, the Ultrahuman Ring AIR's unique feature is its integration with the Ultrahuman M1, helping users attribute blood sugar changes to sleep, stress, and exercise.

It provides "positive multi-dimensional analysis + reverse health advice alerts," allowing precise and targeted glucose control. This differentiating advantage elevates the ring beyond mere health/fitness tracking, imbuing simple biometric data with real medical value.

### 2.6.3 Luna Ring

Noise, established in 2014, set out with the mission to popularize the Internet-age lifestyle among Indians. Initially, Noise focused on developing the mobile accessories market, mainly selling phone cases and accessories. In 2018, Noise started selling smartwatches and TWS Bluetooth earphones. With a business model centered on low cost and low pricing, Noise witnessed a 17-fold growth within just four years and expanded its business to over 8,000 local off-line outlets in India, maintaining a leading position in the smartwatch market for five consecutive quarters.

The Luna Ring, a new wearable device launched by Noise, features an outer ring made of aerospace-grade titanium alloy, with a surface coated in a corrosion-resistant, scratch-resistant diamond-like carbon layer. The inner ring is made from skin-friendly, hypoallergenic material.

This new wearable device is designed to act as a second skin. Catering to various skin types, the Luna Ring features a hypoallergenic smooth inner wall and directional edges on the outer wall to guide the user in wearing it. Embedded within the ring are sensors, including an optical heart rate sensor, a red LED (for blood oxygen and PPG) sensor, a skin temperature sensor, and a 3-axis accelerometer. Noise claims that the Luna Ring can track over 70 biometric signals.

## 2.7 SMART CLOTHING

In the current wearable market, most devices still revolve around being "worn" rather than "worn." Designs primarily focus on smartwatches, fitness bands, smart glasses, smart rings, etc. However, in the "wearable" domain, mature consumer-level wearable products are relatively scarce.

It's important to note that while you can choose not to wear a bracelet or ring, clothing is a necessity, representing a market with rigid demand. In this context, the future might see the development of smart garments that enhance protective functions for humans, potentially making smart clothing the next big thing in wearables. Several smart garments have already emerged, such as the Polar Team Pro sports shirt, Lumo Run smart shorts, and Nike's self-lacing sneakers.

Specifically, the Polar Team Pro sports shirt, designed to suit professional athletes, features a sleeveless design. The shirt has sports sensors sewn into the fabric, which are thin and inconspicuous. Users can access health data anytime, anywhere, without needing to wear wristbands or chest monitors. The Polar Team Pro Shirt incorporates a small sensor with GPS at the back to track the athlete's speed, distance, and acceleration. An iPad app was developed to display real-time data, enabling coaches to monitor each athlete's condition and adjust training plans based on their physical changes.

Since 2010, professional teams from the NBA, NHL, MLB, and NFL have been using the Polar team training system. This system helps coaches evaluate training levels using the collected data, set appropriate off-season training volumes, establish effective performance parameters, and monitor workload during training camps.

Lumo Run smart shorts, developed by Lumo in California, come equipped with sensors that track the movement of the hips and pelvis, gathering valuable data. These shorts can track stride length, steps, and bounce height while running, offering real-time suggestions to enhance overall running performance and prevent injuries. Lumo Run shorts also feature a nine-axis inertial measurement unit and a low-energy Bluetooth module embedded in the waistband. The Bluetooth module collects data to share with a paired smartphone app, which evaluates the data and provides audio feedback on running more effectively and preventing injuries. Even if users leave their phones at home, the device offers post-run analysis by collecting the same data and syncing it once they return home.

Hexoskin is another renowned product in the realm of smart sportswear. The Hexoskin sport vest integrates various biosensors, collecting 42,000 data points per minute. During the day, Hexoskin can measure heart rate, heart rate variability/recovery, steps, calorie burn, and breathing data. At night, it continues to track sleep and the environment, including sleep positions and heart and respiratory activity. All this data is synced to a companion app via Bluetooth or uploaded online for remote coaches to monitor in real-time.

Samsung, the South Korean electronics giant, has also ventured into smart clothing with its project called The Humanfit, aiming to integrate fashion design with relevant technologies. The launch of Smart Suit 40 came alongside three other products: the smart handbag On Bag, which has a built-in battery module for wireless charging of smartphones; Body Compass, a shirt equipped with an ECG sensor to track the wearer's heart rate and breathing; and Perfect Wallet, a wallet and card holder with built-in NFC functionality, which can perform various NFC functions through a smartphone app. When wearing this smart suit, users can utilize a dedicated app and platform to access a variety of useful functions and customize settings according to their needs.

Additionally, the Swedish heated sock brand Seger, in collaboration with the Swedish innovation company Inuheat, introduced the Serger Heat heated socks, optimized for alpine skiing, hiking, or hunting. The battery connects to the socks with a magnetic holder below the cuff, with heat adjustable via a smartphone or directly on the battery, lasting up to 14 hours. Another innovation includes Skiin underwear, capable of recording real-time heart rate, resting heart rate, and heart rate variability.

In 2015, Nike released sneakers with self-lacing technology, which has been upgraded annually since then. These sneakers not only monitor athletic data but also automatically adjust the tightness of the laces based on the wearer's activity. In 2016, Anta, in collaboration with Foxconn, launched its first smart running shoe, the Core Running Shoe, featuring comprehensive running posture monitoring to scientifically test foot inversion, landing method, force, lift height, stride, landing time, etc., helping consumers fully understand their running posture for timely correction and injury prevention.

Beyond fitness and health features, some smart clothing can predict diseases. In 2012, the company First Warning Systems from Reno, Nevada, US, introduced a smart bra for women aimed at estimating the risk of breast cancer in advance. According to 2020 data from the World Health Organization, breast cancer accounted for 11.7% of new cancer cases worldwide (about 2.3 million cases), surpassing lung cancer as the leading type of new cancer globally. This "breast cancer prevention" smart bra uses built-in sensors to detect suspicious lumps in the breast.

Technologically, the bra's principle is straightforward: since cancer cells typically cause abnormal blood vessels, leading to temperature changes in the affected body part, the bra uses subtle temperature sensing and recording, followed by scientific calculations, to predict results. The company's tests showed that this smart bra's accuracy in predicting breast cancer is higher than hospital X-ray examinations, and it can diagnose breast cancer up to six years earlier, providing patients with more time for treatment.

## 2.8 MORE DIVERSE FORMS

Currently, most wearable devices take the form of wristwatches or bracelets, which have not fully tapped the potential of wearable technology. Essentially, wearables are about sensor integration, and devices like smartwatches, bands, glasses, rings, and clothing represent just the initial product forms in this field's evolution.

As new sensors continue to emerge, diversity in form factor will be a major trend for wearables. Some devices may seamlessly integrate with the user's body, becoming a natural part of the person. For instance, researchers at the Institute of Bionic Engineering and Biomechanics at Xi'an Jiaotong University have developed a smart tattoo made from hydrogel microneedle patches for simultaneous monitoring of various health-related biochemical indicators. This smart tattoo mimics the tattooing process, using a soluble microneedle patch as the "tattoo gun" to release color-revealing reagents into the subcutaneous layer for detection.

Another example is non-invasive BCIs, which allow direct interaction with computers or external devices through brainwave signals without traditional physical input devices. This integration enables more intuitive and natural device control. Unlike conventional wearables like watches or glasses, BCIs can deeply integrate into our daily lives, possibly hidden within our hairstyles or clothing, making wearable devices feel less like external appendages and more like parts of our bodies.

Currently, many wearables still function as smartphone accessories, relying on Bluetooth or Wi-Fi for basic information exchange. For example, data from some smart bracelets can only be viewed via a smartphone app, limiting the devices' usability. However, as technology advances, wearables are expected to gain more autonomy, become truly independent, and integrate more closely with other devices and systems. For instance, eSIM smartwatches are becoming increasingly independent, capable of functioning without a smartphone, meeting the communication needs of children and elderly watches. According to Unisoc, the sales proportion of eSIM smartwatches has risen from 17% in 2021 to 23% in Q3 2023. Moreover, with the evolution of Bluetooth and UWB technology, smartwatches are increasingly interconnected with smart cars, smart locks, and other devices.

Looking ahead, we may see even more independent wearables that can integrate with other smart systems. For example, smart clothing might connect with smart home systems, allowing embedded sensors to monitor physiological indicators and activity levels and transmit this data to the smart home system. This collaborative operation could lead to a more intelligent lifestyle, including automatic indoor environment adjustments and personalized health advice. This comprehensive integration not only enhances device utility but also elevates the role of wearables in daily life, making them an integral part of living.

Moreover, the application domains of most current wearable devices are somewhat narrow, focusing mainly on monitoring daily calorie burn, heart rate during exercise, blood pressure, and blood oxygen levels. But wearables have the potential to offer much more. For instance, in September 2022, a research team at Stanford University developed a wearable device that can "monitor tumor size in real-time," enabling timely monitoring of patients' cancer treatment efficacy. With AI breakthroughs, more versatile smart wearables are emerging. For example, in November 2023, Humane, a startup closely watched in Silicon Valley, launched AI Pin, a wearable camera with a laser projector that can display interfaces in the palm of your hand, equipped with an assistant as sharp as ChatGPT. Priced at USD 699 with a USD 24 monthly subscription, AI Pin has caused quite a stir in the tech world, not just for its screenless design but for representing a new concept in human-computer interaction.

Broadly speaking, wearable devices are predominantly focused on consumer markets, but they have the potential to do much more for humanity, bringing unexpected benefits and surprises.

CHAPTER 3

# THE FUTURE OF WEARABLES

## 3.1 FOUR DEVELOPMENTAL STAGES OF WEARABLE DEVICES

In the realm of wearable devices, we can identify four developmental stages:

(1) The first stage is the digitalization of human physiological and behavioral data.
(2) The second stage transforms wearables into control centers for the IoT.
(3) The third stage expands human sensory functions.
(4) The fourth stage involves merging with or replacing human organs.

Currently, the wearable device industry is primarily in the first stage, focusing on digitalizing human physiological and behavioral characteristics. This stage, despite being the initial phase, has undergone extensive development. Wearables were primarily sensor-equipped devices that collected and transmitted human physiological data to smartphones and other smart terminals. Initially, the "intelligence" in wearables was somewhat superficial, serving mainly to provide convenient and timely self-awareness.

However, with advancements in AI, IoT, cloud computing, and big data, wearables have evolved beyond mere data collection. They now offer data analysis, personalized recommendations, and user interaction by leveraging cloud-based storage and analytics. This transition marks a shift from "pseudo-intelligence" to genuine "intelligence," enabling wearables to not only inform us about our current state but also suggest actions or directly facilitate them.

Looking ahead to 2024, the focus will likely remain on deepening the digitalization of human physiological and behavioral characteristics. This focus necessitates further development in both industry technology and talent. This stage will involve extensive exploration and application, particularly in wearable medical devices, potentially reshaping medical technology and practices.

The value of this first stage is immense. Digitalizing human data lays the foundation for all subsequent stages, including the realization of IoT's potential. It also signifies technology's true service to humanity, a challenging but transformative process that involves creating industry chain technologies and data standards.

Despite the challenges, the wearable device market has seen a convergence in product offerings after years of development. Most products are external wearables like smartwatches, bands, and glasses, focusing on fitness and basic health monitoring. From a future perspective, the industry is currently investing heavily while experiencing gradual returns, indicating a phase of significant effort with relatively slow rewards.

As wearable devices evolve and stabilize in their first phase, they will transition into the second phase: becoming control centers for the IoT. In this era of ubiquitous intelligence, when every physical object is equipped with sensors and connected through wearables, these devices will serve as the primary bridge between humans and their environment. In this stage, wearables will play a crucial role, transitioning from a human-centric focus to a central hub for interaction between humans and all connected things. Whether it's for smart homes, smart cities, or everyday activities and travel, wearables will become indispensable intelligent "assistants."

While the current focus of wearables remains in the first phase, some are already exploring the possibilities of the third phase, which expands human sensory functions. This includes using wearables to facilitate communication between people with and without hearing impairments across different languages and nationalities and to enhance human senses like vision, hearing, and taste. However, this phase is still in its exploratory stage and is unlikely to be fully realized in the near term. Once wearables do advance into enhancing human sensory functions, concepts like the mythical abilities to see and hear over great distances could become a reality.

The ultimate phase of wearable devices is their integration or replacement of human organs, propelling humanity into an era of "superhumans." This future is especially likely with the advent and maturation of biological chips, AI technologies based on human brain functions, and the integration of organs with wearable devices, eventually leading to a fusion of humans and robots. Wearables are set to usher in a true "superhuman" era, with future societal structures emerging in ways beyond our current understanding.

## 3.2 THE INTELLIGENT KEY LINKING HUMANS AND OBJECTS

In the IoT era, wearable smart devices represent the unique key that connects people with the physical world around them.

Today, whether it's smart homes or smartphones, current smart terminals are mostly focused on connecting objects to each other, solving the intelligent connection and information relationship between objects. However, if humans want to realize the ultimate idea of smart technology serving people, it's necessary to use wearable devices to establish the connection between objects and people.

In fact, one of the greatest values of wearable devices, which also distinguishes them from smart homes, smart cities, or the IoT, lies in their unique capability as the only device in the mobile Internet era that can carry and realize the connection between humans and smart hardware.

"There's a basis for saying that future wearable technology will replace smartphones as the center of the world." Although current mobile applications and some applications of wearable hardware themselves need to be based on smartphones, fundamentally, smartphones and smart home hardware don't differ much in essence. Smartphones are just a communication tool to which we've attributed more functions. However, the core difference between wearable devices and smartphones lies in the data connection between humans and objects, something smartphones cannot achieve.

Specifically, first, as a medium connecting humans and objects, wearable devices realize real-time monitoring and data collection of individuals. Through wearables, data on an individual's physiology, activity, sleep, etc., can be recorded and transmitted to smart hardware in real-time, providing a rich information basis for personalized services and health management.

In terms of physiological monitoring, wearable devices are typically equipped with various sensors, such as heart rate sensors and blood pressure monitors. These sensors can monitor users' physiological indicators in real-time, providing comprehensive health data. For example, by monitoring heart rate, wearable devices can reflect users' exercise intensity and emotional states and even provide early warnings for potential health issues. This real-time physiological monitoring helps users better understand their body conditions, enabling them to take appropriate health management measures.

In terms of activity monitoring, wearable devices often come with accelerometers and gyroscopes, accurately recording users' movements, steps, activity duration, etc. This provides users with comprehensive activity data, helping to formulate personalized exercise plans. Additionally, by connecting with smartphones or other devices, this data can be synced to apps, forming detailed activity histories for users to access anytime.

Second, wearable devices enable a deep integration of smart technology with individual lives. Whether it's smartwatches, smart glasses, or health monitors, these devices seamlessly connect individuals with smart hardware, becoming an extension of their bodies in daily life. For example, smartwatches not only display time but also integrate various smart functions. By syncing with smartphones, users can receive messages, make calls, check calendars, etc. Additionally, smartwatches often feature activity tracking and heart rate monitoring, facilitating users' exercise management and health monitoring.

This deep integration transforms smartwatches from mere timekeeping tools into essential assistants for daily life and health management. Smart glasses display information directly in the users' field of vision. With smart glasses, users can access real-time navigation, view messages, take photos, etc., without needing to pull out their phones. This changes the way users interact with information, allowing them to safely and conveniently access needed data while walking, driving, and more.

Moreover, a unique aspect of wearable devices is their real-time perception and feedback on user behavior. Through sensing technology, these devices can accurately capture users' dynamic information, enabling more intelligent interactions. This real-time perception and feedback allow smart technology to better adapt to individual needs, providing personalized, customized services.

In essence, only wearable devices can truly integrate with the human body, bind with it, recognize its physical characteristics, and quantify and digitize all these aspects. Hence, in the future, whether it's smart cars, smart homes, smart cities, or the IoT, if they aim to effectively connect with people, they must go through the smart key of wearable devices.

Furthermore, the value of wearable devices lies not only in being a new entry point for the mobile Internet but also in that, in the next wave of business, industries like AI and big data must leverage this smart key to connect with people and derive commercial value by solving their problems. This irreplaceable role will redefine what wearable devices are and how everyone in this field perceives and contemplates the true value of wearables.

PART TWO

# VARIOUS PERSPECTIVES ON WEARABLE BUSINESS MODELS

In the past, the business models for smart hardware products, such as phones, tablets, cameras, and music players, were predominantly based on straightforward hardware sales. Before the Internet revolution, most commercial transactions followed a relatively simple model of exchanging goods for money. However, the Internet has changed these transaction models. Now, we might use Product A for free, but we're indirectly paying for Product B.

In the era of wearable devices, hardware becomes secondary. The pure hardware business model or hardware-plus-app model that we see today is just the initial stage in the development of this field.

In essence, the business model for wearable devices will not just focus on the hardware. When a certain user base is achieved, the ultimate goal is to monetize the traffic and data through analysis and application. This implies that consumer behavior, transaction models, and business models will undergo profound changes. The front end will no longer be the main profit segment; instead, business models extended from the back end will become crucial value points.

Especially in the upcoming Web 3.0 era, which centers on the commodification and valorization of data, wearable devices, with their unique data monitoring and production capabilities, will become a focal point of business competition. In the Web 3.0 era, it is foreseeable that profits from selling wearable devices will not be the main business model. Instead, the primary model will likely revolve around the commercialization and trading of data value generated from the use of these devices.

CHAPTER 4

# SALES OF HARDWARE AND DERIVATIVES

## 4.1 THE FUTURE OF WEARABLES

The development of wearable devices is a long-term trend toward the intelligentization of human society. Especially in recent years, as wearable devices gradually shed the stereotype of being mere "accessories" to smartphones, and with major end-product giants competing in layout and making substantial investments, the market's consumption potential for wearable devices has been further unleashed, leading to an accelerated explosion in growth.

According to IDC Global Wearable Device Market Quarterly Tracking Report, in the third quarter of 2023, the global shipment of wearable devices reached 150 million units, a year-on-year increase of 2.6%. Despite modest growth, this still represents the highest shipment volume for the third quarter since 2021 (table 4-1).

**Table 4-1** Top 5 wearable device companies by shipment volume, market share, and year-over-year growth, q3 2023 (shipments in millions)

| Company | 3Q23 shipments | 3Q23 market share | 3Q22 shipments | 3Q22 market share | Year-over-year growth |
|---|---|---|---|---|---|
| Apple | 29.9 | 20.2 | 40.8 | 28.2% | –26.7% |
| Imagine Marketing | 14.3 | 9.6 | 11.9 | 8.3% | 19.4% |
| Xiaomi | 11.6 | 7.8 | 8.5 | 5.9% | 36.0% |
| Samsung | 10.7 | 7.2 | 11.8 | 8.2% | –9.1% |
| Huawei | 8.5 | 5.7 | 8.9 | 6.2% | –4.4% |
| Others | 73.4 | 49.4 | 62.6 | 43.3% | 17.1% |
| Total | 148.4 | 100.0% | 144.6 | 100.0% | 2.6% |

*Source:* IDC Global Wearable Device Market Quarterly Tracking Report (December 4, 2023)

In the Chinese market, as per IDC China Wearable Device Market Quarterly Tracking Report, the shipment volume of wearable devices in China for the third quarter of 2023 was 34.7 million units, up 7.5% year-on-year, with the market continuing to grow. Among these, smartwatch shipments reached 11.4 million units, up 5.5% year-on-year, with adult smartwatches at 5.59 million units (up 3.9%) and children's smartwatches at 5.8 million units (up 7.2%). The wristband market shipped 3.98 million units, a year-on-year increase of 2.2%, while ear-worn devices reached 19.24 million units, up 9.8% year-on-year (fig. 4-1).

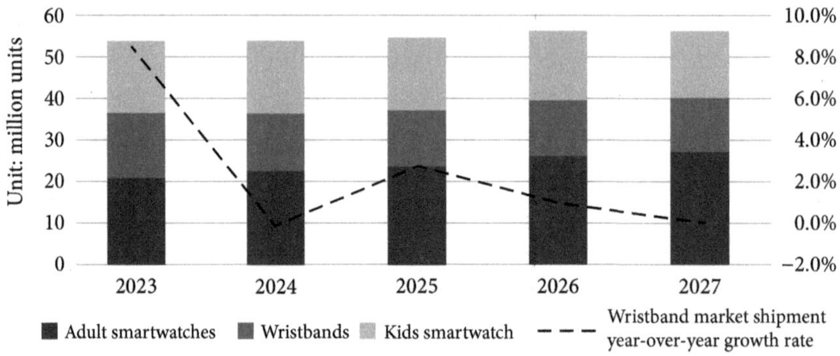

**Figure 4-1** China wristbands market shipment forecast, 2023–2027

*Source:* IDC China (2023)

Looking ahead, wearable devices will remain a hot topic in the tech community. One reason is the strong demand for smart wearables, particularly in health management. As society progresses and the economy develops, humans increasingly demand a higher quality of life and stronger health awareness, shifting from a reactive to a proactive approach in health management. The acceleration of the economy also speeds up people's lifestyles, with high-intensity work causing office workers to easily fall into sub-health conditions, often troubled by chronic issues like cervical and lumbar spine problems or poor blood circulation. According to the 2022 National Health Insight Report, among various age groups, those born after 1995 had the lowest self-assessed health scores, and those born after 2000 experienced the most health concerns over the past year, highlighting the growing health issues among the younger generation.

Furthermore, office workers notably struggle with cervical and lumbar spine issues. In digital office settings, prolonged use of computers and smartphones can lead to shoulder and back pain, stiff necks and lumbar regions, and eye fatigue. The 2021 Office Worker Health Atlas published by the China Business Data Center shows that 67% of office workers suffer from cervical and lumbar spine issues, tied with sleep problems as the most common health concern in the workplace, thus driving the demand for smart wearable devices.

Additionally, the pandemic has heightened consumer focus on personal health, with smart wearables that can monitor key health metrics like heart rate, blood pressure, and sleep being particularly well-received, making home self-testing a new trend in health management. In terms of exercise monitoring, smart wearables offer multi-dimensional data tracking and provide sound exercise advice and management, becoming a "necessity." Core among these devices are smartwatches, which, compared to fitness bands, offer more powerful, comprehensive, and visual functionalities, customizable features for specific verticals, and an aesthetic appeal. The rise of smartwatches is primarily due to advancements in communication and battery life, allowing for independent use and opening up imaginative possibilities for the market.

On the other hand, high-performance and emerging wearable categories have received positive market feedback. In fact, the growth in global wearable shipments for Q3 2023 was mainly driven by the rapid development of small brands and new categories, such as the increasing attention to smaller, more fashionable smart rings. Additionally, new brands like Oura, Noise, BoAT, and Circular are set to introduce related products, bringing fresh perspectives to the wearable market while inspiring established brands in product innovation. Smart glasses, propelled by companies like Meta and Amazon, are also expected to see significant growth.

It is foreseeable that, with changing health concepts and continuous iterations of wearable devices, the future will see an expanding trend in the sales of wearable hardware and its derivative products, positioning it as a new "blue ocean" in the electronics consumer market and a battleground for major manufacturers.

## 4.2 BEYOND TOC: THE TOB OPPORTUNITY

Today, wearable device hardware and its derivatives are thriving in the ToC market. However, compared to the complex and diverse ToC market, the more vertical and specialized ToB commercial market is also an opportunity-rich sector that should not be overlooked. Compared to the diverse and complex ToC market, the ToB market is more vertically specialized and offers at least two advantages.

On the one hand, compared to the multifaceted ToC use cases, ToB applications are more focused, with controlled working environments and repetitive tasks that follow predictable patterns. Hence, the demand for and data from wearables in ToB scenarios are relatively clear, leading to more effective interventions from wearable devices. For instance, a low-risk individual might find monitoring blood sugar via a smartwatch as an unnecessary intervention, creating undue anxiety. In contrast, high-risk diabetic patients might find the smartwatch's monitoring insufficient compared to professional glucose meters, not meeting their expectations. Wearable devices are valuable for health management, and their value varies among individuals. However, this issue is less prevalent in ToB contexts where the needs are more concentrated.

Take sanitation workers as an example. In hot weather, outdoor labor increases the risk of heatstroke, significantly raising the likelihood of severe heat-related illnesses, which can be life-threatening. Traditionally reliant on manual oversight through patrols, identifying health risks among sanitation workers was inefficient and challenging. Smart wearables can collect targeted data to mitigate occupational safety risks in such vertical scenarios.

In Hangzhou, for instance, smart wristbands were distributed to frontline sanitation workers, many of whom are older. These wristbands, customized for their wearers, monitor physiological indicators in real-time. If an abnormal heart rate persists beyond the alert threshold, an automatic warning is sent to the backend system. If the signal is lost or someone stays in one location too long, indicating potential safety risks, the monitoring platform receives an alert. Workers feeling unwell can use a one-touch call feature to dial their team leader's number.

Similarly, scenarios like shoulder and neck strain in desk-bound workers, emotional stress or cardiovascular anomalies in high-intensity workers like IT technicians, occupational hazards like pneumoconiosis, high temperatures, or isolation for miners, and cardiovascular concerns for outdoor workers can all benefit from specialized algorithms in targeted scenarios, clearly demonstrating the product value of smart wearables.

On the other hand, ToB contexts can enhance the intervention effectiveness of smart wearable products more than ToC scenarios. Wearables need to measure and perceive physiological data, encompassing data collection, storage, usage, and management, each with inherent security risks. In ToC settings, such as home health services, many worry that aggressive health devices could compromise privacy, leading to some resistance toward new products. However, in ToB settings, wearables often serve as endpoints in intelligent management systems, with personal health data managed on corporate or industry IT infrastructure. Companies must adhere strictly to national data protection laws, alleviating employee concerns about wearing smart devices and resulting in higher compliance with health management.

Notably, in ToB contexts, wearables are closely integrated with industrial intelligence, involving large quantities and significant demand. Thus, industries and businesses tend to be conservative and cautious in their selections, often partnering with reputable, technologically reliable, and secure giants. As part of an intelligent cloud-edge-end solution, these devices are purchased in comprehensive packages.

## 4.3 WHICH DEVICES ARE PROFITABLE

From the perspective of the business model for wearable devices, although the greatest value of wearable devices does not lie in the hardware itself but in the extensive data generated from being attached to the human body, selling more hardware units remains the quickest and most effective way for most wearable device manufacturers to make a profit. It is also relatively straightforward. Since the focus is on the hardware, manufacturers need to concentrate on the following aspects to make their products more competitive and capture a larger market share:

### 4.3.1 Work on the Design

Each product must possess more than just looks and prove itself to be easy to use. It should be comfortable to wear, with fewer but more effective functions. With the mobile Internet, consumers are increasingly picky. They want more in the products. If a product fails to tick these boxes, it will be thrown to the back of a drawer once the novelty is gone.

This is an era of rapid technological development and differentiated consumers, making the design ever so important. Consumers are not only buying goods for their practical use. They are looking for fashionable looks and something unique. The products they buy need to reflect the taste of their owners and provide sensory and emotional satisfaction. With multiple potential formats available, wearables provide great opportunities for creativity. When design is well done, products will certainly gain consumer attention.

### 4.3.2 Build Killer Applications and Functions

It is, in fact, rather difficult to discover genuine customer needs through market research. If you ask people what they want from their smart devices, the answers are often obscure. They may offer some suggestions, but the true needs could be deeply hidden.

The iPhone has taught us well in this regard. Before the birth of smartphones, users had never asked for a phone with such functions. Steve Jobs first discovered these potential consumer needs and created an iPhone accordingly. It has changed the whole world in terms of communication, socializing, or even our way of life. Apple's Apple Vision Pro, in the integration of AI and BCI interaction control of these hardcore black technologies, in combination with Apple's extreme aesthetic design, can make the product a phenomenal presence.

From the level of business demand acquisition, the actual needs of users are often hidden behind some superficial demands. Only with insightful research and analysis could they be discovered and turned into tangible functions. If a wearable device wants to gain customer recognition quickly, the R&D behind so-called "killer" applications and functions is the key. Nothing else works without this. Meanwhile, killer apps and functions also provide a technological barrier for self-protection against homogenization, allowing a smart device to protect its place in the market for longer.

### 4.3.3 Make Bespoke Products and Provide a Customized Service

We are seeing a return to user-led design. Even industry leaders like Apple and Google cannot rely on their technological edge to ignore consumer needs. They respond to them by introducing different glass frames and colors to cater to customer preferences and maximize market share. Apple has also launched different models with different price ranges in all types of straps based on consumer segments. Everyone likes to be different in some way. A generic product would be difficult to sell when variety is so prevalent.

### 4.3.4 Win with Soft Values

When the Apple Watch was first introduced, many people could not accept the price. Did Apple worry about people being put off by their high prices? It could be argued that this was a deliberate move. What enables Apple to capture the hearts of so many? It is not only because of its hard values but also its soft values. The exterior design of Apple products and good customer experience from an easy and hassle-free system are the main reasons customers cite for their attraction to the brand. In the mobile phone industry, the iPhone is always relatively expensive. Nevertheless, endless consumers are content to pay for them, even if it is really beyond their means.

### 4.3.5 Build Wearables into Luxurious Items

The release of Apple Vision Pro and the earlier Apple Watch exemplifies this approach. The Apple Watch even prompted many traditional fashion watch manufacturers to consider how to make their watches both smart and stylish. Today, smart devices can be considered fashion in and of themselves. If wearable technology can leverage this momentum and step into the luxury goods market by adding value to products through brand effect, premium material, and fine handicrafts, it will surely increase product sales revenue. For example, the sales price of the Apple Watch Edition ranged from RMB 74,800–126,800 (USD 10,535–17,860). Equipped with an 18K gold watch case, this version was visually distinct from the other two versions. The price difference is largely determined by the choice of strap, from the cheapest fluoroelastomer straps to the most expensive modern-style leather straps. This price range obviously is already at the level of luxury wristwatches, which likely still appeals to many customers. When fine workmanship is combined with AI, it turns out to be the most "in" item of the moment. For affluent and trendy people, what could be more attractive?

However, it's important to note that solely focusing on hardware sales won't provide sustainable profits for companies in the wearable device industry, nor will it maximize revenue potential. The future's most promising profit model for wearable devices lies in harnessing big data. In this context, hardware, which is currently the main source of revenue, will eventually become an "accessory product."

CHAPTER 5

# BIG DATA SERVICES

In reality, the true value of every wearable device is not in the device itself but in the big data value it generates, the underlying business models, the ecosystem value created by cross-industry layouts, and the collaborative services provided to customers along with other businesses on the same industrial chain. As the industry matures, the importance of data and services to the wearable sector becomes increasingly evident. Without data, wearables would lose their significance if they only provided location services and interaction with devices like smartphones. Genuine big data services not only bring direct benefits to businesses but also add convenience to the lives of everyday users.

## 5.1 THE POWER OF BIG DATA

Today's business competition has transformed into a competition over data. With the accelerated advancement of the digital economy globally and the rapid development of technologies like 5G, AI, and IoT, data's key strategic resource status in influencing business competition has gained widespread recognition. Dominance in the next wave of global business competition hinges on acquiring and mastering more data resources.

In March 2014, the term "big data" was first incorporated into a government work report, signaling the beginning of its prominence across various sectors in China. In March 2016, the 13th Five-Year Plan explicitly proposed the implementation of a national big data strategy, kickstarting the comprehensive and rapid development of the domestic big data industry. As China's big data industry ecosystem becomes increasingly robust and its integration across various sectors deepens, the national big data strategy enters a phase of intensification. By 2020, data officially became a factor of production, with its market-based allocation elevated to a national strategy, highlighting big data's evolution into a new way of thinking and a hallmark of the era.

Big data, by definition, involves vast amounts of data. Big data technology is a new technological architecture that extracts value from large volumes of data through acquisition, storage, and analysis.

Regarding the volume of data, traditional personal computers handle data in GB/TB quantities. In contrast, big data deals with data volumes at the PB/EB/ZB level. To illustrate, 1TB of storage can hold approximately 200,000 photos or 200,000 MP3 songs, whereas 1PB of big data, requiring about two data cabinets for storage, equates to around 200 million photos or 200 million MP3 songs. 1EB would necessitate approximately 2,000 data cabinets for storage.

The global data volume continues to soar. According to Statista's statistics and forecasts, global data generation was expected to reach 47 ZB in 2020, and by 2035, this figure is projected to hit 2,142 ZB, indicating an impending explosion in global data volumes. The era of big data has truly arrived.

Beyond its sheer size, big data's "bigness" lies in the significant value it delivers. As early as 1980, futurist Alvin Toffler asserted in his book *The Third Wave* that "data is wealth," emphasizing that value is the core essence of big data. The widespread enthusiasm for big data across various sectors, driven by the belief that it enhances competitiveness, stems from the insights and knowledge gleaned from processing and analyzing big data, providing clues and technical foundations for innovative application and business model designs.

For instance, Sesame Credit gathers and aggregates data on individuals from various angles, including identity traits, behavioral preferences, social connections, credit history, and fulfillment capabilities, to assess personal credit. Based on credit ratings, it then develops and operates a range of products, such as credit cycling, convenient transportation, basic communication, credit lending, and recycling.

Furthermore, the value of big data was highlighted during the recent pandemic, particularly in monitoring, tracking, and controlling the spread of the virus. Through government and corporate collaboration, big data products and services were developed to assist decision-making by providing real-time information for government agencies, businesses, and the public. Numerous big data enterprises and Internet platforms utilized their technological advantages to offer services like online education, telemedicine, remote work, contactless delivery, and online entertainment, facilitating digital transformation for many small and micro enterprises.

As a tradable and appreciable commodity, big data represents a fundamental characteristic of the big data era. International data trading began around 2008, with visionary companies increasing their investment in data businesses and exploring new data application models, such as data markets, data banks, and data trading conventions. In China, data trading started around 2010, with the Action Outline for Promoting Big Data Development released in September 2015, explicitly encouraging the cultivation of data trading markets and exploring data derivative product trading to establish comprehensive data resource trading and pricing mechanisms.

In an era where traffic reigns supreme, the underlying data, specifically big data, determines the competitive landscape. At the core of human-engaged big data is user behavior data, constructed by wearable technologies, underscoring the indispensable role wearables play in the data-driven landscape of tomorrow.

## 5.2 WHEN ADVERTISING MEETS BIG DATA

Big data has reinvented the advertising industry overnight by enabling highly effective and precise marketing.

Businesses traditionally spend a great deal on gaining market intelligence. Whether it is knowing who their customers are, what triggers them to buy, where and when they are likely to buy, at what price range, or where they might use the products, the list is endless. With the rise of wearables, we are moving toward a digitalized horizon for all consumer behavior. Businesses and marketing consultants can perform data analytics to determine the most fact-based and accurate marketing ideas. Advertisers can, therefore, impact the target audience most effectively.

To best capture the advantages offered by big data, business models framed around collaboration with social media companies or platforms offer great potential. Potential collaborators include Google, Facebook, Baidu, Tencent, and others. Why is this the best? It's simply because these companies have control over all the information that consumers generate while searching the web, networking, chatting, and emailing. Baidu's search engine index reveals how topical a specific keyword is at the moment and the demographic characteristics of those looking for certain keywords. For instance, the search results of the word "diet" could reveal the type of diet people are more interested in. This information is of great potential value to companies in this business, though it may seem worthless to the individuals searching. By analyzing this information, overall customer preference can be understood. Products can therefore be developed targeting specific customer segments.

Other personal information—including daily interests, food and clothing preferences, and location—used to be hard to gather at a large scale before the arrival of big data. Now Mate or WeChat provides an endless source of data through users' online chatting, sharing, and blogging activities.

The greatest value of search engine companies like Google or Baidu lies in the fact that they retain user search histories on which the logic processing of big data service is based. Gmail publishes targeted advertisements based on content scanning of the user's email. There is a joke on the Internet about receiving advice on how to commit suicide after using the word "suicide" in business emails. In some ways, this could be seen as a violation of privacy when advertising is targeted in this manner. And yet precise marketing also helps meet user needs far better than ever before.

In 2013, Procter & Gamble launched a campaign in China under the name of "Pretty Mom," using search engine data from Baidu. This campaign was awarded the Best Digital Marketing Case of the Year. They had a clear customer group to target: young mothers who were not using disposable diapers. It worked.

Baidu is the number one search engine in China with the best Cloud system as well. It has established the most powerful big data platform in China through its talent-rich big data unit. As Baidu adopts an open platform strategy, it is also the preferred choice for more businesses and brands.

What Baidu offered Procter & Gamble was a convincing statistics-based argument: mothers using disposable diapers have, on average, 37 minutes of additional free time per day compared to those who don't use them. Baidu then further analyzed the activities of mothers during that additional free time and discovered that apart from the expected childcare concerns, they also tended to use the free time to focus on body recovery and pampering themselves after childbirth. This is how Procter & Gamble chose the theme of "Pretty Mom" to launch their online campaign. Sales of Pampers soared in the following season as a result.

Similarly, there are numerous big data marketing case studies on TikTok, such as the one involving Ocean Spray, a cranberry brand established in 1930 in the United States, known for its dried cranberries, cranberry snacks, and cranberry juice.

In September 2020, Ocean Spray's stock price skyrocketed, almost doubling within a few days. This surge began with Nathan Apodaca, an ordinary American uncle living in Idaho, a 37-year-old warehouse manager working in a potato factory, who wasn't well-off and drove a frequently breaking down old pickup truck. One day in September 2020, when his pickup broke down again, Nathan calmly grabbed his skateboard and a bottle of Ocean Spray cranberry juice. While cars zoomed by, he skateboarded, took a selfie with his phone, sipped the cranberry juice, and listened to the classic 1977 song "Dreams" by the British rock band Fleetwood Mac. His expression was serene and intoxicated, occasionally humming along to the tune. This simple and unpretentious video suddenly went viral, garnering tens of millions of views on TikTok within a few days, showcasing the power of algorithms. Driven by big data, the algorithm continuously feeds traffic as the views gain attention, representing a diverse array of users.

In an era where data and traffic reign supreme, the profit model for wearables isn't based solely on the revenue from devices like fitness bands.

Instead, it delves into user big data to enter industries like healthcare, security, education, and advertising, becoming a data provider to offer more precise and personalized services across various sectors. From this perspective, the precise data generated by wearable devices not only describes a person's behavior but can also enable personalized recommendations. Hence, the profit model for wearables isn't just about the limited hardware value at the front end; data and services are the ultimate market battleground for wearable devices in the future. They are also the core carriers of the metaverse and pivotal in constructing the Web 3.0 business ecosystem.

## 5.3 BUILD DATA ANALYSIS MODELS

The business model of data services is a commonality among intelligent terminal products. However, different products will have distinct business models. For instance, the model for a smart fitness band differs from that of a smart range hood.

For wearable devices, intelligent terminals merely collect and monitor various user data. While the functionalities seem extensive and the data appear perfect, users might find themselves at a loss with this information, finding it of little utility. However, if these data are sent to health and medical institutions for analysis, resulting in feedback and health or medical advice for users, a complete service process can be established. Therefore, a comprehensive service solution must back the terminal product to enhance its value.

Specifically, using apps or other means, branded wearables have all adopted Cloud technology to store and exchange data. Analyzing this data could help generate new business models. What wearables provide is a channel for sequential and continuous data collection. Heart rate, blood pressure, and other information are uploaded onto the Cloud from the wearables and synchronized with mobile apps while remaining stored on the Cloud.

The collaboration model between wearables and big data is integrated into the configuration of the big data platform and is needed for making use of preliminary data. Specific measurements of the human body from activity trackers are not of much importance to ordinary users, but they are the inputs for later data processing. Data analysis companies specialized in processing such data have become an emerging trade.

Once integrated with AI systems, the system can instantly recognize, categorize, and process various behavioral data generated by users, forming decision-making suggestions. Alternatively, based on users' behavioral habits and preferences, the algorithm can execute self-decision-making. This application is most widespread in the realm of social media, where platforms utilize user behavior data like viewing duration, likes, and shares to determine content quality and execute targeted recommendations.

The integration of AI provides robust support for the commercialization of big data. For instance, health management enterprises, after establishing relevant data standard models based on medical health standards, can input user data into these models. AI can then analyze specific user data, derive intuitive conclusions, and generate health reports. Enterprises with such AI models can collaborate with health management organizations, insurance companies, and medical institutions to share data results. Wearable device-based health data monitoring can also offer individual health management services. Under AI doctor supervision and management, users can receive solutions and health management advice more swiftly, accurately, and professionally than from human doctors. Health management organizations can use this data to provide health alerts and preventive services to their clients, while insurance companies can adjust premium standards accordingly. Thus, with wearable devices as the data collection entry point and AI as the decision-making system, a complete industry chain is formed.

Based on research on business models of wearable technology in the US, there are insurance companies in America that are already incorporating wearables into their businesses, sometimes building unique business models. We can mainly see two types of models.

In the first type, insurance companies pay wearable producers when their customers use those services, while in the other type, insurance companies encourage healthier lifestyle habits by adjusting premiums according to customer activity records.

The diabetes management company WellDoc is an example of the first type of business. Currently, they are using a "mobile + Cloud" diabetes management platform that focuses mainly on the mobile healthcare side. However, greater value would be possible if wearables were involved in this model.

Currently, WellDoc relies on mobile phones to record and store user blood sugar data to be uploaded to the Cloud. After analysis, it provides personalized feedback to the patient and reminders to doctors and nurses. This system has proven to be clinically effective and economically viable and has been approved by the US Food and Drug Administration (FDA).

The BlueStar app by WellDoc provides timely messaging, guidance, and education services for patients diagnosed with Type II diabetes who are in need of regular drug treatment.

WellDoc charges fees based on the services it provides to users. As this service has been proven effective in reducing long-term costs for healthcare insurance companies, two insurance companies paid over USD 100 per month for clients with diabetes who use this "Diabetes Management System."

The other business model is mainly based on data mining and application. Wearables can monitor and record real-time body statistics through various embedded sensors. Insurance companies could use such data to know their customers in terms of lifestyle and daily activities. To reward healthier lifestyles and exercise, they could lower premiums as a result and increase premiums for those who fail to achieve certain targets.

This is mutually beneficial for insurers and for those who adopt healthier lifestyles. Insurance companies save costs on paying out claims, and customers have more incentive to live healthily. The healthier the customers, the lower the costs.

In the US, most medical insurance is paid jointly by employers and employees. A wearable-associated health insurance scheme could reduce such costs for employers and encourage employees to keep fit and live healthily. This in turn increases workforce productivity. Wearables thus offer many potential returns, killing multiple birds with one stone.

CHAPTER 6

# SYSTEM PLATFORM AND APPLICATION DEVELOPMENT

Clearly, it's challenging for stand-alone wearable devices to generate profits. For wearable devices to truly penetrate the market, they need a value loop that transitions from product to data and then to services.

Based on this, manufacturers of wearable devices aiming for sustainable growth should not just position themselves as producers of smart wearables. Instead, they should be upstream data participants in the larger health industry, medical industry, sports industry, or even the security industry. Wearable device manufacturers should have their own data-sharing platform and be able to open SDK interfaces to create corresponding functional modules. They should also be able to vertically segment data modules according to various application scenarios in sports, medicine, and other industries. This segmentation facilitates more industries and products to access data conveniently in numerous application contexts, enabling the easy utilization of various data types.

## 6.1 MAJOR INTERNATIONAL BIG DATA CLOUD SERVICE PLATFORMS

In the past, the wearable device sector was quite fragmented, reflecting the absence of a unified big data cloud service platform. Many wearable device manufacturers focused solely on producing and selling hardware, only advancing to the hardware plus application stage. This was especially true for startups that lacked the resources to develop a big data analysis platform.

However, the wearable industry has evolved, and following several rounds of consolidation, most wearable device manufacturers have now established their platforms and applications. For instance, on the international tech giant front, Google has Google Fit and Verily, Apple offers the Health Vault and Microsoft Cloud for Healthcare, Microsoft has its Microsoft Cloud for Healthcare, and Amazon provides Amazon Comprehend Medical and Haven Healthcare. Clearly, mobile health platforms have become a new means to attract users and a significant avenue for major tech giants to leverage wearable devices to enter the mobile health and medical market. These giants aim to use their platforms to gather user health data from third-party devices and applications, collaborate with clinics, hospitals, and other medical institutions to enable broader health data sharing and create a standardized health platform to bring more convenience to users.

### 6.1.1 Google: Google Fit and Verily

Google entered the e-health market in 2008 with the launch of Google Health, a health data-sharing service that partnered with CVS Pharmacy and companies like Withings. Users could create a personal data profile on the platform in order to receive healthcare services more conveniently. Google Health failed to incorporate mainstream medical services and insurance companies into the system, and it was shelved in 2011.

Google obviously will not abandon the e-health market. During the Google I/O 2014 conference, they launched Google Fit, which focused on tracking and sharing user activities. In August 2014, Google released the Google Fit Preview SDK to developers. As with Apple, Google Fit also provides an activity-tracking data platform for third-party apps and an API for data storage. In other words, Google Fit abstracts data from third-party activity-tracking apps to generate big data. Its overall strategic direction so far is still focused more on individual users' stats, not support for general medical institutions. If Google Fit were to be connected with Google Glass and other wearables one day, its potential value in the healthcare industry would be very strong.

In addition to Google Fit, another significant endeavor Google has made in medical data service is Verily. Verily is an Alphabet subsidiary that carries most of the healthcare business. This company focuses on improving healthcare through data by using analytical tools, interventions, research, and more.

Verily has established partnerships with numerous healthcare and pharmaceutical companies: it collaborates with medical device manufacturer Dexcom to develop new sensors for tracking blood glucose levels in Type II diabetes patients; it works with ResMed to help diagnose patients with sleep apnea; it assists large pharmaceutical companies like Novartis, Pfizer, and others in modernizing their clinical trials with technology; and it has partnered with Gilead to deploy an immunology analysis platform to gain more insights from the study of biological samples.

In 2017, Verily launched a four-year Baseline Study involving 10,000 participants. This project uses various health tools to collect health data from volunteers, laying the foundation for mapping a comprehensive human health baseline. The concept of a "baseline" is crucial—it's about understanding a person's health benchmark through measurements (via phones, bands, and watches). If one's health improves or deteriorates, the data will show fluctuations, but these are meaningless without knowing the baseline.

Each person's baseline is unique; for example, everyone has a different baseline body temperature. While the average human body temperature is around 36°C, this isn't specific to every individual. For instance, a person whose normal body temperature is 34°C would be considered to have a high fever of 37°C. Therefore, understanding one's baseline is essential for providing personalized and precise health monitoring.

In 2019, the FDA approved an electrocardiogram (ECG) feature on the Verily Study Watch, making it another smartwatch with medical applications.

This watch, also known as the Google Watch, was cleared by the FDA under the 510(k) classification as a Class II medical device. Unlike the non-prescription use authorization granted to the Apple Watch, the Study Watch can only be used with a prescription for patients with cardiac health concerns, allowing for the detection of single-channel ECG rhythms and the recording, storing, and transmission of this data.

The Verily Study Watch features an e-ink display and operates on a platform based on Google's Wear OS, enhanced by DeepMind's AI technology, a fellow Alphabet subsidiary. This positions Google in a vast imaginative space within the wearable medical field.

### 6.1.2 Microsoft: HealthVault and Microsoft Cloud for Healthcare

Microsoft HealthVault was established in 2007. On this platform, with one registered account, a user can set up and maintain their health record. It works like an information safe. It has open interfaces to connect to device makers or insurance companies, but it is up to users to decide who they want to provide that information to.

Users can upload monitored data from other devices to HealthVault, to be analyzed in order to offer users recommendations of potential solutions.

Regrettably, in April 2019, Microsoft announced the closure of HealthVault, an attempt by Microsoft to venture into the Internet-based personal health record system. Despite the shutdown of HealthVault, mobile applications that collect and store personal health information and share it with a user's medical team will continue to revolutionize medical and health practices.

The failure of HealthVault lies in its focus on traditional health records rather than dynamically and in real-time acquiring patients' multimodal data. It lacked integration and interconnectivity with commonly used wearable devices and other smart health devices, and its social sharing capabilities were limited. This restricted its ability to aggregate comprehensive user data, resulting in immature functionalities and user experience. HealthVault did not provide effective information about health issues, such as how a user's health changes over time or what they could do to improve their health, lacking actionable insights and support necessary for users and medical teams to enhance health outcomes and methods.

However, Microsoft did not stop its endeavors in health technology. In October of the same year, Microsoft announced a seven-year strategic partnership with the insurance giant Humana. The collaboration aims to use cloud computing and AI to develop predictive solutions and intelligent automation to support Humana members and their care teams. Humana plans to leverage Microsoft's technological strength, especially its Azure cloud platform, Azure AI, Microsoft 365 collaboration technologies, and Fast Healthcare Interoperability Resources standards, to provide real-time information services to healthcare teams via the cloud platform. By using these technologies to gain a more comprehensive understanding of patients, the goal is to improve preventive care, stay informed about patients' medication plans and supplementary medications, and identify social barriers to health such as food insecurity, loneliness, and social isolation.

In May 2020, during the Build 2020 developers' conference held online, Microsoft highlighted its advancements in cloud computing technologies and deep learning algorithm optimizations. In terms of application tools, Microsoft introduced its first industry-specific cloud solution—Microsoft Cloud for Healthcare. This solution marks Microsoft's first official cloud computing solution tailored for a specific industry, designed to assist doctors and healthcare organizations in providing better services through advanced data analytics, with the advantage of simplifying data sharing across applications.

### 6.1.3 Apple: Fitness App, Researchkit, and CareKit

Apple, leveraging the Apple Watch and iPhone, has amassed a substantial consumer base and obtained vast health-related big data. Particularly, the system-integrated Fitness app on the iPhone offers users a streamlined view to check various types of data, displaying daily fitness records, physical training, medals, and trends in fitness records. It also allows for fitness record sharing and competitions.

In the realm of medical health development tools, Apple collaborates with medical institutions and research organizations to introduce data research tools and product development tools, ResearchKit, and CareKit.

ResearchKit enables researchers and physicians to collect personal data with the consent of Apple users, enhancing the accuracy of their research projects. For instance, the mPower app, developed by the Sage Bionetworks research center through ResearchKit, has recruited over 10,000 participants for a groundbreaking Parkinson's disease research project. The app utilizes the iPhone's gyroscope and other functions to measure participants' agility, balance, gait, and memory, aiding researchers in diagnosing and understanding Parkinson's disease better. Researchers can delve deeper into factors that improve or worsen symptoms, such as sleep, exercise, and mood.

CareKit helps developers build health-related applications, enabling app developers to create applications related to health and allowing users to track health-related data.

CareKit is more like an open-source medical data application platform. It can be used for research, like ResearchKit, and for actual treatment. Some medical institutions apply it to Parkinson's patients to assist with personal care, track medication effectiveness, or for wellness purposes. Symptoms can be tracked via iPhone or Apple Watch, then shared with family members or doctors to help adjust treatment plans.

These functionalities can be displayed or reminded through iPhone or Apple Watch and can connect with other medical health devices. Apple can provide more medical applications through CareKit to aid patients and physicians.

This shows Apple's clear layout in health data services. The Fitness app serves mainly as a personal health adviser, while ResearchKit and CareKit are oriented toward medical research. Furthermore, in terms of the number of applications, the Fitness app will far exceed ResearchKit and CareKit. Although the data from the Fitness app may not always be used for medical research, ResearchKit and CareKit can make significant strides through it.

The future medical industry will gradually be revolutionized in these two directions. By collecting and integrating medical data through health management platforms and then conducting in-depth research and analysis through platforms like ResearchKit and CareKit, precise and effective preventive and diagnostic recommendations can be derived. Subsequently, feedback is provided to medical experts and users, forming an effective medical process loop.

### 6.1.4 Amazon: Amazon Comprehend Medical and Haven Healthcare

In November 2018, Amazon launched Amazon Comprehend Medical, enabling developers to process unstructured medical text and identify patient diagnoses, treatment plans, dosages, symptoms, and signs. It applies AI and machine learning to healthcare, mainly targeting hospital customers to reduce the cost of processing medical documents and to extract information from medical records quickly and accurately.

In April 2019, Amazon announced the launch of six Alexa medical skills compliant with the HIPAA Act, developed by six different medical organizations. For example, members of the pharmacy service organization Express Scripts can check the status of their home delivery prescriptions through Alexa. Cigna members can manage their health through Alexa. Parents and caregivers of children at Boston Children's Hospital can provide the latest information about their children to the care team and receive information about postoperative appointments. Members of the digital health consumer company Livongo can query their blood sugar readings and blood sugar measurement trends and receive personalized data analysis and health tips through Alexa.

Haven Healthcare, launched in 2019 by Amazon's founder Jeff Bezos, Warren Buffett, and JPMorgan Chase CEO Jamie Dimon, is a health insurance service company aiming to simplify insurance and make prescription drugs more affordable. Initially serving 1.2 million employees of Amazon, Berkshire Hathaway, and JPMorgan Chase, it plans to "share our innovations and solutions to help others." Haven Healthcare establishes a doctor network by analyzing performance and cost data, offering essential medical care, including setting up urgent care clinics, integrating medical expert resources, and providing telemedicine services.

Amazon's cloud computing services can assist with the vast data storage and analysis required in the healthcare sector. Amazon's operation centers, supply chain, and acquisition of Whole Foods enable the rapid provision of medical products and services.

## 6.2 THE MAJOR BIG DATA CLOUD SERVICE PLATFORMS IN CHINA

In the 21st century, patients increasingly demand better and more timely access to their medical information while expecting more advanced and affordable medical services. Medical providers face pressures to reduce costs, improve treatment outcomes, increase efficiency, expand services, and comply with evolving regulations. This is especially urgent in medical environments like China, where both patients and providers are eager for improvements in medical care and services.

Technological advancements can help address many of these challenges. New technologies and models can eliminate inefficiencies in business operations or improve the secure sharing of medical information. Therefore, building a medical cloud will become the foundation of intelligent healthcare. The so-called medical cloud is an application form developed specifically for the medical field based on cloud computing. It utilizes cloud computing technology to store and manage medical information systems and data in the cloud, facilitating real-time data access, sharing, and analysis by medical personnel, thus enabling comprehensive management and application of medical information.

Transitioning from cloud computing to a medical cloud involves several essential steps.

First, constructing a cloud infrastructure is necessary. A robust cloud infrastructure is required to support the storage, management, and analysis of medical data, including computing, storage, and network resources while ensuring high availability, performance, security, and scalability.

Second, to integrate medical data into the cloud platform, digitization and management of medical data are needed. Digitization converts medical data from paper or unstructured forms into structured digital data, facilitating storage and analysis. Data management involves categorizing, archiving, backing up, and recovering medical data.

Third, medical institutions can analyze and mine medical data using analytical tools on the cloud computing platform to discover new knowledge and patterns in the medical field. These tools can help physicians better understand patients' conditions, diagnoses, and treatment plans, improving medical efficiency and quality.

Finally, medical institutions can share medical data with other organizations or individuals using services provided by the cloud computing platform. These services can help medical institutions achieve data sharing and collaboration, promoting the exchange of medical knowledge, and improving the quality and efficiency of medical services.

The first cloud service provider in China to lay out a medical cloud was Kingsoft Cloud. In 2015, Kingsoft Cloud and Peking University Medical Information signed a strategic cooperation agreement to jointly launch a medical hybrid cloud solution—"Zhiyi Cloud," which was implemented at Peking University People's Hospital, marking the first step in the layout of the medical cloud.

In July 2018, Kingsoft Cloud released CloudHIS, providing integrated cloud services for grassroots medical workers, residents, and administrators. While connecting resources from higher-level hospitals, Kingsoft CloudHIS supports the use of AI technology and remote medical methods to introduce high-quality medical resources such as doctor groups and expert associations as effective supplements, offering remote medical solutions.

Additionally, Tencent and Donghua Software began a strategic cooperation in various fields, including healthcare, finance, smart cities, and public security, at the end of 2017. Six months after the cooperation with Donghua Software, in May 2018, Tencent invested RMB 1.266 billion in Donghua Software. Donghua Software provides application software development, computer system integration, and information technology services. Combined with Tencent Cloud's capabilities in basic cloud computing services, the two sides have conducted comprehensive and in-depth cooperation in product and solution creation, business channel expansion, project delivery implementation, and operational maintenance services.

In medical informatization, Donghua Software's overall strength is at the forefront of the industry, and it is also a strategic support point for Tencent Cloud's medical informatization solution. In July 2018, Tencent and Donghua Software's wholly-owned subsidiary, Donghua Yiwai, jointly released the "One Chain, Three Clouds" strategy aimed at the medical and health industry:

Health Chain, Public Health Cloud, Medical Cloud, Health Cloud, and six joint solutions.

Following Tencent's investment in Donghua Software, Alibaba also made a move by investing in Winning Health, integrating the big health business system. In fact, Alibaba's strategy started in 2015, when both parties recognized the complementarity in market development and service innovation. They agreed that establishing a long-term, strategic cooperative relationship would facilitate expansion in the health service industry and medical health information services, improve operational space and efficiency, reduce operating costs, and aid in future market expansion. After years of cooperation with Winning Health, Alibaba has formed a medical cloud service chain, including medical management, mobile medical care, medical big data, and pharmaceutical distribution.

Apart from Kingsoft Cloud, Tencent, and Alibaba, other companies like Huawei and JD.com have also initiated actions in the medical cloud business.

CHAPTER 7

# THE OVERLAYING OF BUSINESS MODELS

Wearables are indeed a special type of product. They represent crossovers between traditional consumer electronics and the emerging technology industry, meeting consumers' personalized needs. They offer consumers new functionality to facilitate greater health and convenience in their lives. However, wearable suppliers face challenges in the task of building sustainable businesses in the long term.

Although design has been seen as the key to the success of particular wearables, without a proper business model sales may not take off as desired. Among many potential business models, an overlaying of different models could be adopted as the main business format for the future leaders of the wearables market. Like Apple, there are the Apple Watch and the multiple apps that work with it, as well as Health Vault and Microsoft Cloud for Healthcare in the health space.

## 7.1 BUILDING A STRONGER IOT

If a company wants to succeed in the IoT, it needs to take care of five key aspects: quality hardware, an independent system, application development, a big data cloud service platform, and a social media platform.

Only when these five elements interweave and support each other can a complete IoT ecosystem be built, providing a solid foundation for enterprises to succeed in the IoT field.

First, high-quality hardware is the cornerstone of the IoT system, including sensors, devices, gateways, etc., responsible for collecting and transmitting various data. This hardware must possess high reliability, stability, and security to ensure normal operation in the complex IoT environment. Additionally, the advancement of hardware directly relates to the performance and efficiency of the IoT system. Enterprises need to invest sufficient resources in hardware design and manufacturing to ensure that the provided hardware meets the demands of different scenarios, thus building a reliable and efficient IoT foundation.

Second, an independent system is crucial for the operation of IoT. This system includes the design of the IoT's network architecture, communication protocols, and security mechanisms. A strong and independent system can effectively manage and coordinate a large number of devices, ensuring reliable data transmission and processing. The system's scalability is also crucial to accommodate the continuous expansion of the IoT scale. By establishing an independent system, enterprises can better grasp the core control of IoT, ensuring the system's flexibility and manageability.

The development of applications is key to realizing the true value of IoT. In the IoT era, applications cover various industries and fields, including smart homes, industrial automation, healthcare, etc. Enterprises need to invest significant resources in application development to meet the needs of different areas. Application development also needs to closely integrate with the characteristics of hardware and systems to maximize their potential. At the same time, enterprises need to focus on user experience, ensuring applications provide users with convenient and intelligent services and promoting the widespread application of IoT technology.

A big data cloud service platform is an essential foundation supporting IoT operations. Through the cloud service platform, enterprises can store, analyze, and process vast amounts of data, extracting valuable information. Big data analysis can provide profound insights for enterprises, aiding in smarter decision-making. The cloud service platform can also achieve remote device management and upgrading, enhancing the maintainability of the entire IoT system. A robust big data cloud service platform is key for enterprises to fully unleash the potential of IoT.

Last, social platforms act as a bridge connecting individuals, enterprises, and devices in the IoT era. Analysis, establishment, and feedback after data collection are critical to building user stickiness. Once users lack stickiness, they are unlikely to continue showing interest. Conversely, if the social circle is bustling, it further motivates users to gravitate toward more lively places. Thus, social interaction is not the goal but an inevitable outcome. Through social platforms, users can share and access information related to IoT, forming a vast community. The presence of social platforms fosters interaction and collaboration among users, aiding in the innovation of IoT technology. Enterprises can establish a brand image, promote products, and gather user feedback through social platforms, better meeting market demands.

This shows that technology giants already venturing into this field, such as Apple, Google, and Samsung, are all deeply cultivating and planning in these five areas.

These world-class tech giants are certainly ambitious when they are not short of money or people, and the impetus they play for the coming era of the IoT will be remembered by history. The era of IoT is no longer simply a battle of hardware but a battle of data and a battle of platforms, just like the two camps of iOS and Android in the era of smartphones.

These IT giants, as early IoT explorers with both financial and human resources at hand, are full of ambition. What they have done so far has been to contribute to the development of this industry, but inevitably, some will win, and some will lose as a result of the risks that early experiments entail. The real battle of the IoT age is not about devices alone but rather about the killer weapons: the data and the platform. Just as with iOS versus Android, the battle of the operation systems is, in fact, the core competition splitting the smartphone makers into two groups rather than the devices themselves.

Google has not been a major supplier of smartphones, but it has contributed to the Android system for many. In the IoT era, Google has gone straight into wearables, smart home devices, self-driving cars, and many other areas. The Android Wear platform they developed is the first platform released for smartwatches. It really shows how determined Google is to exploit all that the IoT has to offer. I believe the next move with system platforms will be to increase openness and integration. The self-developed systems from Apple, Google, Microsoft, and Samsung will eventually be able to connect and exchange data in the background.

Tech giants benefit from their leading positions in this industry, as they are able to implement many different business models simultaneously. This is a major barrier for tech startups, who simply cannot afford the costs. We can see many startups in China and overseas using crowdfunding to finance initial product development. Lack of investment funds forces these companies to focus on device sales to gain funds sooner. Most of these products are relatively simple in terms of technology and production chain. These intelligence-light products avoid the higher production costs of more complex technology. Product failure is something these companies cannot afford.

Therefore, startups generally adopt the business model of creating a popular product to market in return for quick wins and rapid returns on investment, or for further financing possibilities. Their contributions to the industry are limited to the sale of hardware devices. They improve the appearance and design details to make the devices more attractive and comfortable, or they upgrade parts of the devices to improve battery life or make data acquisition more accurate. These startups struggle to get involved in big data analysis or platform building. They simply do not have the capacity or infrastructure to support such activities. Instead, these companies are more likely to focus on device-making alone and seek to be connected to big data platforms provided by others. Or they could choose to use the data for research or analysis by collaborating with medical institutions, insurance companies, or other third parties.

## 7.2 THE GREAT OPPORTUNITY OF A FOCUSED MODEL

As previously noted, the most profitable business model may be a combination of co-existing business models. However, this is not a strategy fit for all businesses. Most startups would not be able to adopt such a model. Industry leaders, on the other hand, are best suited to employ this strategy, as they have the necessary funds to fuel the complex structures and the human resources needed. It is still better for the startups to choose the stand-alone device model. They could follow this route and explore opportunities in the following areas:

(1) Using market segmentation to build system platforms. System platforms can be built according to vertical market segmentation based on product types, including activity trackers, smart clothing, smart footwear, and more. Such platforms are less technically challenging, relatively, and easier to optimize around focused targets.
(2) Building technologies for the production chain. Technologies used in nodes along the production chain can be used as entry points, such as voice interaction, batteries, smart materials, sensors, chips, and so on. Such technology-focused business models can create competitiveness, as long as the development team is equipped with sufficient technical resources and capacity.
(3) Providing application-led solutions. Focusing on one portion of the production chain, these companies may look at providing apps, algorithms, technical solutions, or manufacturing for related technology startups, including designing crowdfunding schemes, with the aim of building unique strengths to provide solutions in a specific area.

As the entire wearable technology industry grows rapidly, technology and products are constantly upgraded. All parts of the production chain are gaining momentum and presenting even more opportunities, so choosing any entry point may be a great choice for aspiring entrepreneurs looking to start businesses with wearable technologies. In particular, smart sensors are the cornerstone of the meta-universe era, the era of wearable devices. This kind of sensor that is digitally monitored by the user is no longer the single, traditional data monitoring of the past, but the smart sensor that has self-calculating and decision-making functions only after it is integrated into the AI chip. In the foreseeable future, smart sensors will show explosive growth, from the digitization of things and the digitization of the environment to the digitization of people; the core of the era of the digitization of everything lies in the smart sensor.

PART THREE

# WEARABLE BUSINESS APPLICATIONS

Wearable technology is still in its infancy. Most business models still focus on the development and sales of hardware and accessories. As the industry develops, an ecosystem will establish itself, providing support for many potential business formats. Free from reliance on hardware sales as the sole profit-making channel, Big data-based wearables will trigger new business opportunities in healthcare, tourism, education, gaming, fitness, advertising, public life, and many other areas. This part of the book examines the potential developments in a range of sectors. As we look ahead, we can imagine the potential for radically different business models—models of which businesses today should be conscious and the potential gaps in the market that can be exploited.

CHAPTER 8

# WEARABLE + HEALTHCARE

The potential for wearable technology in hospitals, health-related areas, and other medical applications is immense. This may include the use of sensors in providing care and using the latest wireless and Bluetooth technologies to collect real-time measurements from different parts of the body to enable better treatment, whether preventive or when intervention is needed. Medical care could be transformed by the use of different sensors, with machine learning-generated algorithms supporting medical staff in the process of diagnosis and treatment. Wearable technology may allow for early intervention before a patient is even aware of their ailment. Wearables can also assist in delivering healthy lifestyles—in terms of diet or exercise—and provide medical staff with data on users' activities and other measurements of interest.

The use of systems approaches, based on huge new platforms, is needed to support the use of wearable technology in this area. There are a number of key components of a system that need to be in place to ensure the health benefits of wearables are fully captured. To date, most wearables—including smartwatches, wristbands, clothing, and footwear—are only used to monitor user body readings for the purpose of fitness management. This could help address a number of public health issues, from obesity to cardiovascular disease. However, this technology has not yet been integrated fully into medicine. As we look ahead, there is a plethora of potential applications of wearables in the medical area—

fitness management is but an early step in tackling the health concerns facing the planet in the current century.

## 8.1 RETHINKING HEALTH MANAGEMENT

Medical services serve health. With the support of modern medicine, human life expectancy continues to increase. Modern medicine has created new scenarios, changing the relationship between humans and themselves, with diseases, suffering, and death, and also changing people's definition of "health."

Of course, there are as many definitions of health as there are people. In the past, health was often closely associated with "illness"—a person who was not sick was considered healthy. The "progress" of modern medicine has brought more scientific and advanced disease detection methods, which can even predict diseases before they occur.

This has somewhat led to the "generalization" of diseases. Everything can "get sick," either now or in the future. At the same time, the improvement of living standards and the influence of the recent consumption upgrade trend have also quietly changed society's dimension of health assessment. The understanding of health is no longer as it used to be.

In 2020, a landmark review published in *Cell* detailed the eight core markers and dimensions of health, including spatial compartmentalization (barrier integrity and local interference containment), homeostasis maintenance (recycling and renewal, system integration, and rhythmic oscillations), and appropriate stress response (homeostatic resilience, xenobiotic excitation effect regulation, and repair and regeneration), providing a systemic new definition of health from the perspectives of the whole organism, organs, cells, sub-cells, and molecules.

Spatial compartmentalization is divided into barrier integrity and containment of local changes. Barrier integrity, apart from skin, intestines, and respiratory tracts providing barriers against the external environment, includes various barriers at different scales within the human body. These barriers form important electrophysiological and chemical gradients and facilitate the exchange of gases and osmotic pressure, replenishment of metabolic circuits, communication/coordination between compartments, and detoxification. The integrity of these barriers is crucial for maintaining health. For example, the blood-brain barrier, formed by tight junctions of various cells in the neurovascular unit, limits bacteria or inflammation-causing chemicals in the bloodstream from entering brain tissue. "Leakage" in the blood-brain barrier is found to be related to various neurological diseases.

Containing local changes refers to the body's response and repair to minor local changes, including external injuries, pathogen invasions, various "accidents" during cell division leading to DNA repair failures, accumulation of erroneous proteins, etc. This includes barrier healing, self-limiting inflammation, innate and acquired immunity, anti-tumor immune escape, etc. By promptly controlling minor local disturbances, overall health is achieved.

Maintaining homeostasis is divided into recycling and renewal, system integration, and rhythmic oscillations. Recycling and renewal refer to the need for most cell components and cell types to constantly enter the cycle of death, clearance, and renewal to avoid degeneration. Maintaining a healthy organism involves the "integration" of different systems, from intracellular structures to organs and tissues, to the interaction between the human body and the microbiome. Networks at different levels are intertwined, with many elements playing several roles simultaneously.

Moreover, the precise order and timing control of molecules and cells during embryonic development or regeneration are vital for life. Ultradian, circadian, and infradian oscillations provide rhythmicity to physiological functions, aiding in maintaining homeostasis. Irregular rhythmic oscillations, such as frequent late nights, disrupt homeostasis, leading to health issues.

Last, an appropriate stress response is closely related to homeostasis, xenobiotic excitation effect regulation, and repair and regeneration. The body relies on internal stability to be relatively independent of external conditions, thus enhancing its tolerance range for ecological factors. Homeostatic loops maintain countless biological parameters, such as blood pH, serum osmolarity, arterial blood oxygen and carbon dioxide, blood glucose, blood pressure, body temperature, body weight, or hormone concentrations, at near-constant levels. If the set point of the regulator is changed, it will lead to chronic diseases.

Hormesis refers to the phenomenon where exposure to low doses of a toxin can induce a protective response to prevent damage from higher doses of the same toxin. For various health-threatening damages, repair is essential. These repairs involve DNA and protein molecules, as well as organelles like the endoplasmic reticulum, mitochondria, and lysosomes. Where possible, damaged or lost functional components need to regenerate for full recovery.

It's evident that health is not merely about the absence of disease. Establishing a new concept of health based on modern medical standards is essential for contemporary healthy living. With this new perspective, the concept of "health management" has rapidly evolved.

Although people today have a new understanding and consensus about "health," many are still unfamiliar with health management. Simply put, health management is the comprehensive management and focus on an individual's or group's health, aimed at preventing diseases, promoting health, improving quality of life, reducing medical expenses, enhancing health literacy, and achieving longevity through scientific means.

Health management encompasses prevention, medical care, rehabilitation, and more, marking a shift from the traditional disease-centered medical model to a prevention-focused new model. It's an emerging discipline that intersects medicine, public health, and health sciences, holding significant importance for individual and societal health.

The evolution of the health management industry dates back to the 1970s, initially focusing on disease control, public health, and medical quality. With the increasing focus on health since the 1990s, the industry has shifted toward individualization, especially with the support of wearable devices, big data, and AI.

In 2009, the US government introduced the Health Information Technology for Economic and Clinical Health Act to promote health information technology application and development. This act encouraged medical institutions and doctors to use electronic health records to improve medical efficiency, reduce costs, and enhance quality. Subsequently, the health management industry saw new technologies and applications, such as telemedicine, mobile health, and health monitoring devices.

In 2010, Apple launched its first smartwatch, Apple Watch, becoming a leader in the wearable device industry. Its release marked the beginning of the smart health management era, enabling people to more easily track and manage their health data.

In 2015, China's State Council issued Opinions on Promoting the Development of the Health Service Industry, setting the policy goal of "strengthening health management and preventive health services." Since then, the health management industry in China has developed rapidly, becoming an important sector in China's medical and health field.

Today, with the continuous development and application of wearable devices, AI, and big data, the health management industry is facing new development opportunities. According to iResearch, China's health management market grew from RMB 45 billion in 2014 to RMB 185.8 billion in 2020, with an annual compound growth rate of 29.7%, and is expected to exceed RMB 300 billion by 2022. Grand View Research forecasts that the global health management market will reach USD 308.8 billion by 2028, with an annual compound growth rate of 22.3%.

As an emerging industry, the development of health management in recent years is unprecedented. In the foreseeable future, health management will become an essential component of modern medicine, helping individuals or groups prevent diseases, promote health, improve quality of life, reduce medical expenses, enhance health literacy, and achieve longevity.

## 8.2 WEARABLES: DATAFYING HEALTH MANAGEMENT

The development of the health management industry is inextricably linked to the continuous advancement and popularization of modern technology, among which wearable devices hold a special significance. As an intelligent key bridging humans and objects, wearable devices have truly opened the door to health management, bringing a significant transformation to the entire health and medical field. The primary value of wearable devices lies in digitalizing the physiological and health-related characteristics of the human body, distinguishing them from any other smart product. Unlike smart homes, smartphones, or robots that only provide external intelligence, wearable devices adapt based on the individual's physiological changes. Especially for mobile medical products, if they are solely based on smartphones without deeply integrating with an individual's physiological characteristics, the issues they address remain superficial, like appointment scheduling or payment processes.

Thus, wearable devices are not just about miniaturizing smart hardware; their real value lies in digitalizing people's dynamic and static behaviors and physiological characteristics. This transformation is not only revolutionizing human life and commerce but is also genuinely enabling mobile healthcare and health management.

Among wearable devices, smartwatches are the most representative. On September 10, 2014, during Apple's fall event, Cook unveiled Apple's first smartwatch, positioning it as a sports and health device. Specifically, the Apple Watch is equipped with a heart rate sensor, accelerometer, gyroscope, and barometer, which are capable of monitoring heart rate, tracking calories burned, and providing a health data report.

Following its launch, smartwatches spread like wildfire in the health and medical fields. Just nine months after the Apple Watch hit the market in 2015, its shipments reached 11.6 million, surpassing the total annual shipments of smartwatches in 2014, which were under 7 million. Over the next few years, the Apple Watch made significant strides, even surpassing Rolex in 2017 to become the highest-selling watch globally.

From a demand perspective, health functions have become one of the most influential factors for consumers when purchasing smartwatches, and smartwatches are playing an increasingly vital role in the healthcare sector. A 2021 survey by Global Market Monitor showed that among the many features of smartwatches, health monitoring is far more important than calling, video, or positioning features, with over 70% of potential consumers prioritizing comprehensive health monitoring capabilities when choosing a smartwatch.

On the one hand, this is because modern individuals are increasingly focused on their health, particularly as cardiovascular diseases, the leading health threat in China, are becoming more prevalent among younger people. News of sudden deaths among the young continuously underscores the importance of health awareness. Not only young people but also the elderly are showing a growing demand for smartwatches. For instance, if an elderly person falls or experiences a sudden health anomaly, the smartwatch can immediately contact emergency contacts or even automatically send an alarm, ensuring their health and safety while providing peace of mind for their families.

On the other hand, the trend toward device intelligence necessitates an efficient key for controlling these smart devices. Smartwatches, which combine a fashionable tech appearance with the ability to provide real-time data on exercise, sleep, heart rate, blood oxygen levels, and more, offer a great solution for modern individuals keen on monitoring their health metrics.

Taking blood oxygen monitoring as an example, a blood oxygen saturation below 94% is considered insufficient oxygen supply. Many clinical diseases cause insufficient oxygen supply, directly affecting the normal metabolism of cells, making blood oxygen detection crucial for clinical medicine. However, the most primitive method of measuring blood oxygen required blood collection followed by electrochemical analysis using a blood gas analyzer, which was complex and did not allow for continuous monitoring.

With the evolution of clinical medicine, non-invasive blood oxygen measurement is now commonly used. By equipping patients with a photoplethysmography sensor, continuous blood oxygen monitoring can be achieved. This method uses red light with a wavelength of 660 nm and near-infrared light of 940 nm as light sources, measuring the light transmittance through the tissue to calculate blood oxygen concentration and saturation, displayed by the device. Similarly, smartwatches can measure blood oxygen by determining arterial blood oxygen saturation and assessing the wearer's health status.

Moreover, such features are abundant, including assisting rescue teams in locating individuals who have fallen off cliffs or providing blood oxygen reference values during COVID-19 prevention and control. Leading tech companies are vigorously researching and optimizing the health monitoring functions of smartwatches. At the end of 2021, Huawei launched its first smartwatch capable of measuring blood pressure, Huawei WATCH D. Apple Watch's ECG and atrial fibrillation notification features were introduced in China, with constant rumors about its upcoming blood sugar and blood pressure monitoring functionalities. Additionally, companies like Huawei and OPPO have established sports and health science laboratories to overcome technical challenges in the field of sports health.

Intelligent wearable devices' health monitoring capabilities are virtually irreplaceable, signaling a long-term trend in health management. In the future, wearable devices, like today's smartphones, will completely transform lifestyles. For example, during morning exercises, shoes could calculate the distance and calories burned, glasses could capture the scenery, and Bluetooth headphones could monitor blood oxygen levels. Wearable technology is poised to permeate everyday life, bringing about significant technological transformations.

A decade ago, few anticipated that smartphones would replace computers as the primary Internet access device for all ages. Similarly, today, few believe that wearable devices could become the next smartphone, transforming lifestyles and offering significant investment opportunities for the next decade.

Wearable devices, serving as new portals for mobile networks, will lead to a comprehensive upgrade of personal area networks. Their appeal lies in liberating humans from the constraints of computers and smartphones, creating new mobile network entry points. Current mobile networks, heavily reliant on smartphones for server and input/output functions, face limitations. The widespread adoption of wearable devices will change this scenario, relegating smartphones to server roles while wearables become the primary mobile network interfaces, allowing hands-free Internet access anytime, anywhere.

"It is foreseeable that wearable devices will significantly reduce overall medical costs," said Simon Segars, CEO of ARM, a UK-based company. People in remote areas can send high-definition data from home and receive remote analysis and treatment, avoiding the hassle of travel. Combining wearable devices with the Internet, big data platforms, cloud computing, and professional medical services will streamline the medical process and offer unprecedented comprehensive health management.

In the future, wearable devices will continuously track various health metrics, recorded and analyzed through the Internet, cloud, and mobile health platforms, providing targeted health advice and recommendations for medical resources, if needed. This new era of wearable devices will not be confined to a narrow scope but will integrate seamlessly into all aspects of life, delivering precise and abundant information far beyond traditional Internet access methods.

As discussed, wearable devices, narrowly defined as sensor wearables, offer users insights into their health status through data generated by sensors distributed across the body. With in-depth data analysis, users can gain accurate insights into their health and make timely adjustments, independent of medical professionals.

For instance, if someone with hypertension or heart disease is nearing their alcohol limit, the wearable device could issue a warning, suggest dietary adjustments, and prevent further drinking. If someone falls ill without knowing the specific ailment, the wearable device, observed by countless virtual doctors, can quickly provide a health report and prescription, which is then shared with partnered pharmacies for prompt delivery.

In summary, future medical care will significantly reduce overall costs, especially time expenses for patients, addressing a major flaw in traditional healthcare. Everyone will be able to easily monitor their health, becoming their health manager, with doctors playing a supportive role.

## 8.3 WEARABLES AND CHRONIC DISEASE MANAGEMENT

Over the past 200 years, human life expectancy has more than doubled, a monumental achievement largely attributed to advancements in modern medicine and public health initiatives that have enabled more people to survive childhood diseases and extend their lifespan.

However, with the intensifying trend of an aging population and the increasing number of elderly individuals, the rise in chronic diseases has become a prominent issue. Particularly, chronic diseases such as diabetes, Parkinson's, and Alzheimer's, which have subtle early symptoms that are hard to detect, require substantial human and material resources for daily care and nursing once they reach an advanced stage, severely affecting the patient's health and quality of life.

In the United States, about 60% of adults suffer from one or more chronic diseases, ranging from heart disease and asthma to Alzheimer's, kidney disease, and diabetes. This imposes a heavy burden on the healthcare system, which struggles to provide adequate services, and the cost of managing these diseases is high. In the US alone, nearly three-quarters of healthcare expenditures are related to chronic diseases or associated complications.

In China, there are also over 300 million people with chronic diseases, and deaths caused by chronic conditions account for 80% of all disease-related deaths in the country. The cost of managing chronic diseases accounts for 70% of the total national disease expenses, becoming a significant public health issue affecting the country's economic and social development.

The increase in chronic diseases is partly due to the natural decline in organ function with age, leading to conditions like hypertension, diabetes, tumors, and cardiovascular diseases. Additionally, changes in modern lifestyles have also negatively impacted the prevalence of chronic diseases, such as poor dietary habits, lack of exercise, and high stress. Moreover, as modern medicine evolves, more chronic diseases are being effectively controlled, allowing people to live longer. While this reflects the success of medical advancements, it also signifies an increase in chronic conditions.

Managing chronic diseases in an era of increased longevity has become an unavoidable reality, and wearables are poised to be the optimal solution for this challenge. Using advancements in wearable technology to tackle these massive societal problems will allow for the realization of just a fraction of the ultimate value of the technology itself.

For most existing wearables, it takes time for users to get in the habit of wearing them. When a company brings a new technology to the market, building customer awareness can involve incredibly expensive marketing campaigns. In addition, fitness wearables have needed to evolve to meet changing consumer preferences to avoid becoming outdated. The use of interactive social media tools to gamify the use of wearables can help change behavior. Time is needed before users fully integrate the novel device into their lives—and become comfortable with the concepts of tracking and measuring their body's performance. Many tech startups struggle to meet the costs associated with building and maintaining their customer base. Deep pockets are needed to survive in this market.

The situation is quite different for medical wearables, particularly in their use for the management of chronic diseases. Before wearables became popularized, devices like electronic blood pressure meters and blood glucose monitors were being used to improve the daily lives of those with certain medical conditions. Smart wearable blood pressure meters or smart blood glucose monitors can offer new functionality, but they are based on similar concepts of self-monitoring. They may look different from the old devices, but they may also contain vastly improved technology—including the potential for linking to other devices and access to one's own data.

Patients (or consumers) understand what blood pressure monitors do, so it is far easier for users to adopt these than wearables, where the functionality may be less familiar or understood. Such understanding makes the uptake of the technology more likely and probably allows for quicker diffusion of the technology in the marketplace. For companies, the fact that patient or consumer awareness is high shortens the "go-to-market" phase and provides significant cost savings.

An outstanding feature of NCD patients is that their demand is highly focused on the technology's quality, rather than style or customer service. They are unlikely to abandon a device when the novelty factor wears off or because they dislike some element of its appearance. As long as the device delivers the required functionality, NCD patients will be satisfied.

For example, patients with hypertension need to measure their blood pressure and take medication regularly. Their need is for a monitor that can measure blood pressure accurately. Wearable device developers therefore focus largely on making an accurate and easy-to-use device that attaches to the patient's body to allow automatic blood pressure monitoring. The device is also designed to transmit data to the patient's smartphone, which can be viewed in an app. The app may also provide diet and lifestyle advice to help the patient maintain stable blood pressure. The data could also be shared with clinicians in order to reduce the need for consultation or to enable early intervention as needed.

The elderly can also benefit, particularly from smartwatches and smart clothing which measure heart rate, temperature, posture, movement, and other factors. Dr. James Amor, a research fellow at the University of Warwick, says activity monitoring allows families and caretakers to see the elderly person's health and routine. Through the deployment of wearables, community healthcare services could add to digital health records for all residents, which would help to better understand the impact of chronic diseases in the community. The data collected could also be used to assist in disease management and in medical research.

NCD patients will likely become reliable consumers of medical wearables. They are also the users who truly need such technology. In the context of an aging society, with significant burdens from chronic disease, medical wearables may open doorways to new solutions and so improve lives for patients and reduce costs for healthcare providers.

## 8.4 MEDICAL WEARABLES FOR DIFFERENT POPULATIONS

Based on the segmentation, marketing plans could be more specifically targeted for various products, including trackers, smartwatches, smart glasses, and others. It is important to segment the market according to user groups. For example, infants and toddlers, children, women, elderly, or disabled people all present different requirements for the wearables they need. These are discussed below in turn.

### 8.4.1 Wearables for Infants and Toddlers

As this user group is very young and requires special care around the clock, it is important that wearables for them meet stringent safety standards.

Wearables can offer a range of functions to support the care of this group, including monitoring sleep quality, movement, body temperature, heart rate, and other health indicators. Parents can access data and view analysis from a PC or a smartphone. In the case of accidents, such as an infant crawling or falling out of their bed, alerts could be issued immediately to their caregivers.

Quite a number of wearable devices targeting infants and toddlers have developed. They may be a great help to new parents in the first years of parenting. Among the technology in this category is a smart ankle bracelet from the company Sproutling. This monitors the baby's heart rate, sleep position, and whether they are asleep.

The product is made up of three modules: sensors, a band, and a smartphone app. The band is made of soft medical-grade silicone with a lovely red heart design in the middle, embedded batteries, and four sensors. Sproutling also established a health database for babies under 12 months old. Parents can configure the baby's profile by adding their age, body weight, body height, and other measurements to connect to the device through a smartphone. When irregular results are detected, an alert is given to the parents through the app. This may give reassurance to parents, particularly after the child has been moved to their own room.

For instance, a heart-wrenching statistic from 2018 revealed that 2.5 million newborns globally passed away within their first month of life, with babies from Africa and South Asia accounting for 87.7% of these deaths. This alarming figure moved designer Chris Barnes and his team at the UK-based Cambridge Consultants, prompting them to take action and create a remarkable health monitoring device named "Little I." True to its name, like a caring little eye, it constantly watches over the infants' well-being, bringing a ray of hope to newborns in regions with scarce medical resources.

"Little I" is a wearable device shaped like a shoelace, using silicone straps and an ABS shell to secure the electronic components. When placed on an infant's foot, it automatically turns on and begins continuous monitoring of the baby's temperature and blood oxygen saturation using built-in temperature and SPO2 sensors. Its user interface is incredibly straightforward, resembling a traffic light system that uses colors, icons, and unique sounds to clearly convey the baby's health status to the parents. If there's an issue, "Little I" emits an alert to prompt parents to take immediate action.

This wearable device, affectionately known as "Little I," supports infants through their first month. If any problems arise during this period, parents receive timely notifications. After a month, "Little I" is collected postdoctoral examination, then cleaned, charged, and prepared for the next user. By introducing this product, there has been a significant reduction in the risk of infant mortality during the first month.

Similar products now available on the market include a smart sock developed by Owlet, smart baby bodysuits by Mimo and Exmovere, and smart diapers by Pixie Scientific.

All in all, the top four needs for infant wearables to be successful are safety, comfort, accuracy, and responsiveness. Those enterprises wishing to enter this market need to have quality hardware as well as a well-established service platform. Those traditionally in the baby product industry in China have a strong advantage in their knowledge of the market and existing customer base. It is possible that these companies will launch wearable products for babies first, in order to take a significant portion of the pie offered by this market segment.

### 8.4.2 Wearables for Children

Statistics show up to 200,000 children go missing in China every year, and only 0.1% of these were recovered. A missing child has a significant impact on any family. As a consequence, child safety is now an increasingly important public security matter in China.

Most children's wearables on the market today have very simple functions, mainly GPS and location tracking. Undeniably, this is another major battlefield for wearables with strong market demand. Most of the existing child safety wearables use a 3-in-1 operations model: hardware + software + Cloud. Wearable manufacturers have developed not only the hardware device, but also the mobile app and the data analytics platform. A well-designed combination of these can provide a better user experience and more significant returns.

For example, in February 2022, a company in China launched a smart shoe for children. This smart shoe is embedded with a chip that enables location tracking, preventing children from getting lost. Unlike other location-based products that use GPS signals, this shoe, known as the 58°C smart anti-lost shoe for children, achieves real-time location tracking through the transmission of software and hardware signal bands. It has an effective lifespan of up to two years without the need for charging. The chip is small, waterproof, and implanted at the bottom of the shoe, allowing the shoe to be washed without affecting its functionality. Each child's anti-lost chip corresponds to a unique security code. Guardians can bind the shoe to their phone by downloading an app, registering, and linking to the shoe's security code, which can only be linked to one phone number. However, guardians can authorize multiple individuals to monitor the child through the app's sharing feature.

Child abduction is a persistent social issue not only in China but also in other countries. Smart shoes with anti-lost features can potentially deter such crimes. The smart shoe can trigger an alert on the guardian's phone if the shoe moves beyond a preset safe distance, effectively preventing children from getting lost in crowded places like airports, train stations, and parks. Besides the anti-lost function, the smart shoe also includes a water immersion sensing feature. If the child falls into the water, the app immediately sends out an alert, buying crucial time for rescue efforts.

Intelligent wearable products for children meet a global demand among parents. Although there is a significant market need, companies or entrepreneurial teams looking to enter this field must not only develop wearable devices for children with longer battery life, more precise location tracking, and low radiation but also find solutions to ensure these devices remain on the children, addressing the critical issue of devices becoming detached from the child.

### 8.4.3 Wearables for the Elderly

The aging population—and their associated healthcare issues—is a global problem. The cost of care is rising. To address this, wearable developers could look into developing a system of community care based on big data analytics to support care in a person's own home.

China is a country with a strong cultural focus on honoring and caring for the elderly. Looking ahead, smart homecare solutions may become mainstream solutions to caring for the elderly. According to statistics, 90% of older people would prefer to stay in their own homes, with only 10% choosing assisted living facilities. Given that preference, extending care services to those living in their own homes is critical.

The rise of people with Alzheimer's or dementia adds significant pressure to care systems. People with this disease partially lose their ability to look after themselves, struggling to remember things, losing spatial awareness, and easily becoming lost. For this particular market, wearables must have a locator function on top of sensors for heart rate, breathing, and other health indicators.

This type of device is still in the early stages of development. We have not yet seen any major products appearing on the market that specifically target this segment. The main formats of potential wearables are shoes, mobile phones, and accessories. CMA800BK, a credit card size Comfort Zone product recommended by the Alzheimer's Association, weighs only 50 g. Placed in the patient's pocket, caretakers could access a patient's precise location with one click. When the patient steps outside an established "Comfort Zone," caregivers receive text alerts immediately.

Utilizing Qualcomm's inGeo platform, an embedded GPS chip, and an Internet connection, it can be remotely controlled. The current sales price of the product is USD 99.99, but there is also a monthly service charge of USD 14.99. If the elderly patient is able to carry a smartphone, using a Sprint smartphone with built-in Comfort Zone functions achieves the same functionality with a monthly charge of only USD 9.99.

Many of those with Alzheimer's disease have the tendency to wander. They can be stubborn and refuse to accept change, including the use of new wearables. It would normally take a while before the patient is used to wearing or carrying the device with them. To address this issue, GTX and the shoe maker Aetrex have jointly developed a GPS-based shoe product, so the locator can be worn without the person noticing.

GPS-enabled shoes are no different in appearance than ordinary shoes. The embedded GPS sensor allows caretakers to locate the patients through smartphones or computers. The same "Comfort Zone" alert function is available too.

We have seen some companies in China using smartwatches to target health and care services for the elderly. Using the smartwatch, a connection is made between the user, the hospital, and the family. Apart from daily health monitoring and reminders, it could be used during an emergency to call for medical attention.

### 8.4.4 Wearables for the Obese

Today, we are entering an era that prolifically fosters obesity. Over the past few decades, the proportion of obese individuals has been rising worldwide, with particularly severe increases in developing countries.

A 2016 global adult weight survey report published by the renowned British medical journal *The Lancet* revealed that the global adult obese population had surpassed the number of underweight individuals. China has exceeded the United States, with nearly 90 million obese people, including 43.2 million men and 46.4 million women, making it the country with the largest obese population in the world. A study by the Public Health Center at Peking University also indicated that by 2030, one in four children will be overweight. By then, China will have 50 million children classified as overweight or obese.

WHO's latest report also shows that over 3.4 million adults die from cardiovascular disease, cancer, diabetes, and arthritis caused by obesity. It is clear that obesity is a matter of life and death. Yet, so far, no country has managed to find an effective solution.

In solving the obesity problem, fad diets or liposuction are not permanent solutions. Only through regular exercise, healthier diets, and better living habits can this issue be properly addressed.

The real issue when targeting this segment of the market is how to drive this group of people to change their behavior in order to form healthy habits.

Wearables could help in bringing about behavior change. They've been shown to be one of the most popular product types in the diet and fitness market. Wearables have four unique strengths:

(1) Some wearables can be worn 24 hours a day. This contrasts with smartphones, which are usually turned off or moved further away from users during the night.
(2) Wearables can monitor real-time health data around the clock. This is the greatest benefit of wearable devices. The data generated could be applied in many ways to impact daily life, particularly in the field of medicine. Wearables could form the basis of a new era of disease prevention and management.
(3) Social communities formed around wearables and their connection with health insurance may encourage users to adopt them. If wearables are the social norm or health insurance costs are reduced through their use, they may encourage more people to exercise regularly and have healthy lifestyles. The current trajectory of the market for fitness devices and apps is one of high growth—as more people wear them, they will increasingly become the "norm."

As I have repeatedly stressed, the healthcare industry is one of the first growth points in the wearable market. Based on market segmentation, it is likely that eager dieters will be another stronghold of demand for wearable technology.

It is clear that both the market background and the strengths of wearables have enabled wearable technology to be well-exploited in the fitness market. I therefore would like to encourage investors and entrepreneurs to consider these segments as potential entry points into this very promising market.

### 8.4.5 Wearables for People with Disabilities

Many people around the world have devoted their lives to helping those with disabilities live more fulfilling and socially integrated lives. Wearables offer great potential to foster integration and reduce barriers for people with disabilities. It is very likely that through specially developed wearables, people with disabilities could live more independent lives. There is great market potential and room for development in this area, for example, using an exoskeleton to help paralyzed patients stand, using special glasses to regain vision for the blind, or helping the deaf to hear with advanced equipment and technology. Many science and technology firms have been developing and commercializing these products. In this area, the most attention is paid to wearable devices based on BCI technology.

For instance, BrainCo is dedicated to applying BCI core technologies across various fields to create groundbreaking products. These include BrainRobotics' intelligent bionic hand, the StarKids BCI system for autism spectrum disorder intervention, and Mobius bionic legs, among others. The BrainRobotics intelligent bionic hand, a high-tech prosthetic that merges BCI with AI algorithms, can interpret the neural signals from the arm muscles of its wearer, allowing individuals with upper limb amputations to control the prosthetic hand as if it were their own, achieving intuitive movement. In 2019, the BrainRobotics bionic hand was named one of the year's top 100 best inventions by *Time* magazine. In 2020, it won the Red Dot Award: Best of the Best for its design.

Another example is the development by researchers from Harvard University and Boston University in 2024 of a soft, wearable robotic suit designed to assist Parkinson's disease patients in walking without stiffness. This mechanical garment, worn around the hips and thighs, gently propels the hips during leg swings to facilitate longer strides. This device was able to completely eliminate stiffness in indoor walking trials, enabling participants to walk faster and farther than without the garment. The research team noted that their robotic suit, by providing minimal mechanical assistance, could produce immediate positive effects and continuously improve walking capabilities under various conditions. This innovation not only restores mobility for Parkinson's patients but also their independence.

### ▸ ARTIFICIAL EYES

A biotech startup has developed a novel artificial eye product. It uses a system called EYE (Enhance Your Eye) to regain sight for the blind. The company uses

3D bioprinting technology to make artificial human organs. They have so far produced ears, blood vessels, and kidneys. As the head of the company explains, due to the complexity of the human eye, it is not easy to print an eyeball.

To date, the company has managed to offer three different models of the EYE system. "Heal" is the standard version with an electronic iris. The "Enhance" model adds an electronic retina and camera filters (retro, mono, and others). The version with the highest specifications is called "Advance," and it is even equipped with Wi-Fi. Eyes are now electronic devices just like smartphones.

To use EYE, the patient needs to remove the original eyeball and implant what is known as "the Deck," which connects the eye to the brain.

The developers explain that the product will not be available on the market until 2027, therefore no actual photos are available yet.

However, visual impairment has always been a major problem plaguing human society, whether it is congenital visual impairment or visual impairment that occurs with aging. With the combination of intelligent technology and BCI technology, it is possible to regain the ability to see, which can be said to be a very important and revolutionary technology for human society.

### › TENSION CONTROL DEVICES FOR AUTISTIC CHILDREN

Autistic children could face tremendous challenges when under pressure and can struggle to voice their needs. Parents and teachers must be extra careful not to stress these children.

Wearable manufacturers have so far launched two products, Neumitra and Affectiva, to address this problem for autistic children by monitoring the physiological reactions of the user. These devices can also be used for medical purposes by monitoring post-trauma anxiety. These smart bands could benefit thousands in the autistic population, making it easier for their caregivers to track their tension levels and provide support.

Previously, the relevant organizations have begun to test the Affectiva wristbands. According to the Autism Society of Ohio, school teachers will distribute the wristbands to their classes, and then the teacher will use the bands to detect the student's behavior or actions and will be able to determine the student's mental state, including moments of anxiety and relaxation.

### ➤ MIND-CONTROLLED WHEELCHAIRS

On November 19, 2022, the journal *iScience* under *Cell* press published a study demonstrating that, after extensive training, individuals with quadriplegia can maneuver a thought-controlled wheelchair in natural and complex environments. This research is primarily based on BCI technology. "We found that mutual learning between the user and the BCI algorithm is crucial for the successful operation of such a wheelchair," said José del R. Millán, the study's corresponding author and a professor at the University of Texas at Austin. "Our study highlights a potential path for improving the clinical translation of non-invasive BCI technology."

Millán and his colleagues conducted a longitudinal study with three participants with quadriplegia. Each participant underwent three training sessions per week for a duration of two to five months. Wearing a cap without brims, participants' brain activity was monitored using electroencephalography (EEG) and translated into mechanical commands for the wheelchair via a BCI device. The participants were instructed to control the wheelchair's direction by thinking about moving parts of their bodies. In essence, they utilized BCI technology to control the wheelchair, allowing it to navigate through complex environments based on the users' intentions.

While this research has shown progress, some challenges still need to be addressed before its real-world application. Nevertheless, this study illustrates the profound impact that BCI-based wearable devices can have on our society.

### ➤ EMOTIV INSIGHT BRAINWEAR

Previously, Philips, a major European electronics company, collaborated with the Irish consulting firm Accenture to develop a solution that utilizes brainwaves to control home appliances, aiming to facilitate a more normal life for individuals with paralysis. They named this wearable brainwave tracking device the Emotiv Insight Brainware, designed to enable patients with ALS, also known as Lou Gehrig's disease, to control their environment.

The goal of Philips' Digital Accelerator Lab was to integrate the newly developed brainwave controller with other devices, such as its Lifeline Medical Alert Service, allowing users to make emergency medical calls simply by thinking. Furthermore, the Emotiv Insight Brainware scans the patient's EEG brainwaves to create a BCI which then transmits the collected data to a tablet. This allows ALS patients to issue commands through the tablet to control Philips' electronic products, such as adjusting the brightness of lights, turning the TV on or off, and changing the volume.

### › DOT BRAILLE WATCH

A Korean startup team, Dot, has developed the world's first braille smartwatch called "Dot." Like other smartwatches, it can be worn on the wrist. However, instead of a conventional touchscreen, it features a series of raised dots.

The watch can pair with a smartphone, and when a text message is received on the phone, it is translated into braille and sent to the Dot, which alerts the user through vibration. The dots are arranged in four horizontal rows, with six dots per row, capable of moving up and down to display braille characters. The speed of the braille changes can be adjusted and set as needed, with the fastest being 100 characters per second and the slowest one character per second.

In addition to this, Dot also has features like a watch, alarm, and notifications. Blind individuals can simply touch it to know the time, and by pressing a button on the side, the watch will verbally announce the time. Users can also use it to send messages. The device has an open API, allowing any developer to create additional applications to enhance the product.

Clearly, once integrated with AI and leveraging AIGC's generative voice functionality, this device could allow blind individuals to move beyond dependence on braille for recognition and instead utilize AI voice capabilities for auditory playback and interactive communication.

### › PROFESSIONAL SPORTS

Distinct from general fitness wearables, smart devices for professional sports require more precise measurements of heart rate, breathing, and other metrics to monitor an athlete's performance in terms of speed, distance, and endurance. Enhanced data processing software can translate this data to allow sports doctors to understand the athlete's body condition and inform bespoke training or recovery programs. Coaches could also use the information to understand their teams in order to pick the players best suited for the match.

These products have not received much attention from non-professionals. The smart wearable products include clothing and equipment specially designed for sailing, mountain climbing, boxing, and other sports. These products are different from other fitness products because they are able to deliver targeted professional training programs. They can also be used to identify mistakes during training, and so to correct these.

When Zlatan Ibrahimović playing for Paris Saint-Germain, took off his top after a match, it revealed a dark garment shaped like a sports bra underneath, which sparked curiosity among fans. In fact, this was a smart sports vest made by GPSports which was used for tracking a player's body condition and movements in real-time.

Apart from Paris Saint-Germain, Real Madrid, Chelsea, and many other top football teams in Europe have also started to use GPSports products. These so-called "Man Bras" are kept well hidden most of the time.

Other similar wearables include Bro, a rugby top with an embedded data recording device. This product is also a GPS tracker, to help coaches and doctors better understand player movements. When body condition or fitness levels are reduced, coaches can see direct visual data analysis on an iPad or PC, which could enable them to choose players in better condition to play.

Obviously, with the accelerated realization of AI, as well as the accelerated development of intelligent and miniaturized sensors, wearable devices will continue to penetrate into all aspects of our lives, and ultimately realize the era of intelligent wearable of everything between people and things, which lays the foundation for the realization of the meta-universe.

CHAPTER 9

# WEARABLE + TOURISM

## 9.1 MAKE SMART TRAVEL PERSONAL

A truly memorable journey will often require a great deal of time devoted to the planning phase. We can consider the case of budget travel, and unpack some of the key factors that need to be considered in preparing for the trip. These include the following: the overall itinerary, necessary transport arrangements (tickets or rentals), tour booking, checking the weather, researching the destination to plan where to visit, budgeting, arranging documents (e.g., passports and visas), packing suitcases (including clothing for different weather conditions, cameras, chargers, battery packs, and more), and learning some of the languages if traveling to a foreign destination.

Planning a great holiday can be a lot of hard work. How might the tourism industry look in the future with the increased use of wearable technology? Picture this in your mind. One day, you are slumped on the couch trying to find a new TV series to binge, but you can't find anything interesting. You're wearing a garment with subtle sensors that pick up your emotions. Combining this with your social media activity, your bank balance, and your diary in an algorithm based on your preferences, the smart system is able to suggest that perhaps it's time to recharge your batteries with a vacation. Subsequently, these smart wearable devices will, based on our hobbies, tastes, consumption habits, financial capacity, and other factors, utilize AI to formulate and recommend travel itineraries, introduce the features of each route, and the experiences we might gain. Additionally, they will plan the timing of each itinerary based on our work schedules, along with the relevant budgeting, essentially providing a travel plan even more thoughtful than a personal assistant. In essence, the only decisions we need to make are whether to go on a trip and which travel plan to choose.

Wearable technology has already brought profound changes to the healthcare, gaming, and fitness sectors. Such changes will inevitably extend to more areas, including travel and tourism. The Internet has made planning trips easier through better information. Travel in the wearable age will be easier still. The devices would filter information for you, giving you access to a better travel experience.

How could wearables deal with some of the most common issues in travel and tourism? The next section discusses some opportunities.

## 9.2 TROUBLE-FREE SOLUTIONS

### 9.2.1 Identity Recognition

The need to prove your identity is becoming more and more common. Passports, ID cards, and bank cards are indispensable these days when traveling. You need ID documents and cards with various levels of security. Technology has advanced rapidly with facial recognition and other security measures that are implemented in airports and hotels. The use of biometrics is becoming more common—using fingerprints or other metrics for identification. Encrypted biometrics will likely be mainstreamed as the preferred way of verifying a person's identity to give access to social media accounts, smart devices, hotel rooms, and more.

However, for ensuring absolute security in such scenarios, wearable devices emerge as the ultimate choice. Why is that? Because wearing one serves as a form of authentication. Compared to other smart devices, wearable devices are not only the most familiar to the user—since their primary function is to collect data from the user—but also, this data, after subsequent processing and feedback, becomes a unique identity verification code. More importantly, wearable devices integrated with biometric recognition capabilities possess unique identification features, making them more secure, reliable, effective, and unique than any form of ID.

In other words, the identity recognition method crafted using wearable devices isn't solely based on a single biometric characteristic of the human body. Instead, it's based on a combination of specific or abstract data, including heart rate, blood pressure, blood lipid levels, facial features, skin characteristics, and even personal preferences, to create a unique identity code. This identity code is unique and irreplaceable. Even if the device is lost and found by someone else, it won't leak any personal information because its identification and related functions' private data become invalid the moment it's separated from the user's body. It might still show time, but the identity recognition and associated features will be disabled. This is where the immense charm of wearable devices lies, marking their killer application in the realm of technology.

The unique identity code established by wearable devices for users can validate one's identity more effectively than an ID card. A single wearable device can replace an assortment of documents such as passports, boarding passes, and ID cards. In the future, presenting a particular wearable device—or even an implanted chip or electronic tattoo—will simplify complex identity verification processes through a simple scan or sensor detection. When considering the likelihood of loss, which is more prone to being misplaced: various documents or a smartwatch worn on the wrist or an electronic chip tattooed on the body? Clearly, it's the former. And even if a smartwatch is lost, there are still smart bands, smart jewelry, or even smart clothing—any of these wearable items can serve as our identity proof.

Moreover, the payment feature of wearable devices can eliminate the need to carry various bank and membership cards, along with their layered encryption complexities. These devices can also function as hotel room keys or smart car keys. In essence, all card-type and key-type items can be integrated into a single wearable device, and all ticket checks, payment gates, and access controls can be navigated by simply presenting a wearable device worn on a part of the body.

Furthermore, wearable devices can offer reminders and suggestions at appropriate times based on our schedule and backend data on traffic conditions, tourist site status, hotel statuses, etc., aligned with our regular lifestyle. Of course, we can also interact with the device using voice commands to give feedback, allowing the wearable device to directly assist us with hailing a ride, processing boarding passes, buying tickets, or booking hotels.

### 9.2.2 Language Barriers

International travel involves a lot of communication, some of which can be made difficult by language barriers. In some areas of the world, a local guide may be needed. Juggling phrasebooks, tourists could struggle with the basics of ordering food or accommodations, particularly in remote areas. Language poses a barrier when accessing medical care or dealing with the police. Interpretation, when available, may prove to be beyond the means of the average tourist. The need for a device that can act as a universal translator is clear in this context, and some wearable devices are starting to offer a real-time translation function.

The latest smart glasses, such as the Solos AirGo3, are now capable of providing real-time language translation services. Using the SolosTranslate platform, these glasses help break down language barriers and facilitate better conversations between people from different linguistic backgrounds. The SolosTranslate platform, supported by Solos' proprietary software and OpenAI's ChatGPT, is set to transform international business, travel, and cultural exchanges by promoting fast, efficient, and inclusive communication.

The operation modes include a listen mode, tailored for one-on-one interactions, capturing and translating an individual's speech directly in front of the user and discreetly playing the translated speech through the smart glasses in the user's preferred language. This ensures a private and intuitive experience.

The group mode is designed for multi-person discussions, allowing each participant to speak and listen in their preferred language. Users can easily join or initiate group discussions via QR code or web link, enhancing the versatility and inclusiveness of various user preferences.

For text mode, SolosTranslate provides translated messages in written form, enabling users to display translated text for reading or choose to play the translated speech audio. In presentation mode, it allows speakers wearing

AirGo3 to deliver content in their language and have it translated in real-time, enabling the audience to understand in their chosen language.

Solos President Kenny Cheung stated, "AirGo3 is designed to continually innovate and adapt to consumer needs. Part of this need is a reimagined mobile platform that enables us to communicate digitally in real-time in a more natural, hands-free manner."

In the future, as AI advances in voice detection, input analysis, and more, tackling different accents, a multitude of dialects, and changing environments, real-time translation will have even greater development potential. Once we overcome various linguistic and communicative barriers that might affect the travel experience, the next step is to explore how wearable devices can enhance the enjoyment of traveling.

## 9.3 BE YOUR OWN TRAVEL AGENT AND TOUR GUIDE

Wearable devices offer great potential not only for data input but also for output. They could provide an almost endless range of possibilities to the tourism and travel industry. One of the potential impacts is a transition away from traditional travel agents and tour guides.

Traditional travel agents and tour guides may provide information, plan tours, and arrange accommodation and meals. They have a fairly central role in the traditional tourism model, particularly in certain countries where tourism may be less organized or for particular market segments who may prefer a more organized travel experience. How might wearables impact this part of the tourism sector?

Google Glass can project virtual images and provide navigation based on voice recognition, essentially becoming a hands-free tour guide. With the potential for AR, it could revolutionize the traditional tourism market.

In the future, simply wearing AI-powered smart glasses during travel can greatly enhance the experience, ensuring that even solo travelers can thoroughly enjoy their journey. With such technology, there's no worry about getting lost, missing out on authentic local delicacies, taking the wrong transport, failing to book a cost-effective hotel, or visiting a site without understanding its historical context. Essentially, AI can provide all the information you might want to know about a place, while smart glasses can present the entire locale "in plain sight," offering insights beyond what traditional guides can provide.

Travel data packages will become a core element of travel in the era of wearable devices, supporting a new industry and creating jobs. For example, considering the city of Beijing, we can develop various data packages like Beijing travel guides, authentic local cuisine, landmark explanations, and virtual travel experiences. These packages can be sold to companies providing AI travel services or used to train specialized AI travel service applications. Before visiting Beijing, travelers can download relevant data packages to get a preview of the Forbidden City, evaluate travel plans, or discover local delicacies directly as if using Doraemon's "Anywhere Door." If travelers don't want to spend time researching, they can simply communicate their preferences to an AI, which then provides visualized options through smart glasses.

In the future, companies will develop data packages for different aspects of travel, which will be used to train AI models for the tourism industry. Users can easily access these AI applications, simplifying travel planning and decision-making.

Furthermore, these tourism models will be updated in real-time with new data. For instance, a Beijing local cuisine data package within a tourism AI model would rank dishes based on real-time user reviews, not paid promotions. If a traveler finds a dish too sweet, they can leave feedback via their smart glasses, and this input will be integrated into the AI model for future travelers' reference.

When these comprehensive tourism models are combined with AI-powered real-time translation features, travelers can explore the world without language barriers. Travel in the era of wearable devices transcends mere sightseeing, offering immersive and fun experiences deeply integrated into local life. This brings us back to the question: in this new era, what will be the role of traditional tour guides?

## 9.4 TRAVEL LIGHT

I don't know about you, but for me packing luggage is one of my life's frustrations. No matter how well you think you have packed, it can still be a struggle to close the case. And I always end up forgetting something important, even if I've packed the proverbial kitchen sink. From clothing to cameras, the sheer bulk of stuff needed can tax even the best packers. However, wearable technology could transform what you need in your suitcase.

Consider a trip around East Asia, from the snowy mountains of Northern China to the sultry conditions of Singapore and Malaysia. Traditionally, you would have had to pack clothes for the different climates—thick clothes for the cold areas and lightweight ones for the more humid tropics. Smart clothing offers a lot of potential for this kind of scenario. A piece of clothing could self-adjust to the local conditions, providing warmth in cold areas and cooling in hot areas. 4D printing technology offers the potential to create clothing that shifts shape under different environmental conditions, meaning one garment can suit a range of settings.

Capturing videos and pictures is part of the regular tourist experience, but how many of us have struggled through piles of vacation photographs from relatives or colleagues and are unable to truly share the experience? Even with the best cameras, you can miss the essence of the experience. Perhaps the emerging field of VR recording may offer some potential here. This would involve the use of a device to record the experience and share that experience through immersive VR.

Simply by putting on a VR headset, friends and family can experience the same sights and sounds. This may become the new way of recording future travel experiences, capturing the memorable moments of life to be relived again and again. You could use Oculus Rift, for example, to record your African holiday. Sharing that experience back home with your family and friends would be awesome, wouldn't it? Of course, Apple's Apple Vision Pro comes with more powerful features and a more realistic and comfortable virtual visual experience.

Maybe one day in the future, all we need to take will be a pair of smart sunglasses, a device that could be used as a camera, a video recorder, and a VR headset. Recording and sharing fun holiday experiences could be ever so easy. It would preserve your travel experience with more fidelity than ever. It could completely recreate scenarios that are ready to be shared and potentially lived by others.

## 9.5 VR TOURISM FOR THE FUTURE

The so-called VR travel is actually the simplest way to understand that we can travel around the world without leaving our homes.

So far, the application of VR technology has largely been in the field of gaming. In the future, however, this technology will have a far wider impact on many industries, including medicine, education, and tourism. Tourism, especially, could be radically changed. The impact will be on both the marketing of tourism and the way consumers enjoy tourism.

The traditional marketing of tourist destinations is based on photos, videos, and other imagery to convince consumers that the destination is a fun place to be and that the choice to visit is one that the potential tourist will not regret. Yet, most of these methods do not have very much impact. VR technology has the potential to disrupt the tourism market and radically transform the marketing methods used. Consumers would be able to have a genuine taste of the travel destinations through immersive VR experiences. Various VR-compatible videos could then be shot to promote certain destinations, targeting specific segments of the market.

Destination BC, a state-owned tour company for British Columbia, is one of the first companies to use VR technology in a tangible way to promote tourism. They employed Oculus Rift technology to create a VR video called "The Wild within VR Experience." The destination-marketing video was filmed using a custom rig—built with a 3D printer, no less. Seven specialized HD cameras were mounted around the rig, allowing footage to be filmed by helicopter, by boat, by drone, or on foot. The breathtaking views seen in the video were shot throughout the Great Bear Rainforest in the beautiful countryside of British Columbia, Canada.

At the end of 2014, Marriott International unveiled a new travel booth where hotel guests can explore the black sand beaches of Hawaii or the top of London's Tower 42 using the Oculus Rift.

Although sightseeing and touring could be key applications of VR technology in the future, they do not mean that people will no longer want to travel in person. However, using this technology could make travel planning and scheduling much easier, taking advantage of VR preview and demonstration functions.

Marsha Walden, CEO of Destination BC, explained: "We think VR is a great fit for tourism marketing." The initial application for VR technology in tourism would be marketing, as it offers an immersive preview for users. However, the other applications may include using VR to experience the entire journey later on.

## 9.6 THE METAVERSE MAKES TRAVEL MORE MAGICAL

Metaverse brings a new dimension to travel, emphasizing the idea of "living elsewhere." People travel to see different landscapes, experience varied lifestyles, and broaden their horizons. The metaverse can recreate and replicate real-world tourist spots holographically, allowing you to explore the world without leaving home. For instance, you might marvel at the architectural beauty of the Forbidden City in the morning, enjoy the poetic scenery of West Lake in the afternoon, and observe wildlife in Africa at night. Compared to traditional photos, videos, and live streams, the metaverse offers a full-sensory, immersive experience, enhancing the perception of nature and culture through VR.

With the metaverse built on wearable VR devices, not only is the immersive travel experience heightened, but it's also more convenient than physical travel. There's no need to worry about crowded tourist spots; just prepare the necessary equipment and explore the world from your couch. The metaverse not only opens doors to virtual travel experiences online but also integrates with physical tourist spots to innovate and diversify the tourism industry. For example, virtual guides could replace traditional signposts, and AR can bring historical performances to life in a meadow, enhancing visitor engagement.

Furthermore, metaverse projects can transcend time and space, offering unique opportunities for less renowned destinations to attract tourists and generate revenue through immersive, game-like experiences. These projects could include virtual agriculture, fishing, or craft-making activities, promoting local industries.

Metaverse travel, built on VR wearable devices, can also offer historical recreations or futuristic adventures, allowing users to showcase "superpowers" in a virtual setting. For those with limited time and budget, metaverse travel offers a comprehensive and authentic way to explore destinations like Xinjiang, experiencing its geography, culture, and cuisine through immersive VR and data from other travelers.

It is particularly great for destinations some of us may never be able to reach or which are threatened by tourism congestion, like the ruins of Pompeii, the interior of the Pyramids of Giza, or many of the ancient Chinese landmarks. You could even visit the surface of Mars or the bottom of the sea. VR technology could make the impossible possible.

CHAPTER 10

# WEARABLE + EDUCATION

Things can be done much faster and more efficiently today thanks to mobile Internet. Every aspect of our lives has been transformed significantly, including education. In recent years, the Internet + education as a representative of education greatly impacted the education industry, trying to break through the disadvantages of the traditional way of education. Especially in China, where educational resources are scarce and unevenly distributed, and the cost of education is high, this change has become even more urgent.

## 10.1 THE RISE OF INTERNET + EDUCATION

Over the past decade, the rapid development of the Internet has profoundly and broadly impacted education, transforming traditional educational paradigms and providing learners with more flexible and diverse ways of learning. This transformation is evident not only in the updates to academic content and teaching methods but also in the globalization of educational resources, the rise of online learning platforms, and the personalization of educational models.

First, the Internet has promoted the globalization of educational resources. Previously, learners primarily depended on local schools or libraries for educational resources, limiting their access to information. Now, the Internet breaks geographic barriers, allowing learners to access a wealth of educational resources worldwide, including Massive Open Online Courses (MOOCs), digital libraries, and academic paper databases. This globalization of educational resources broadens and deepens learners' access to the latest knowledge and research findings.

Second, the Internet has spurred the rise of online learning platforms. With the advancement of Internet technology, numerous online learning platforms, such as Coursera, edX, and Udacity, have emerged. These platforms offer a variety of online courses and degree programs, providing learners with more flexible learning opportunities. Learners can choose suitable courses based on their interests and needs and arrange their learning time autonomously, transcending temporal and spatial limitations. This flexibility and convenience make education more aligned with individual learners' needs and encourage active and engaged learning.

Moreover, the Internet has facilitated the development of personalized education models. Traditional education often adopts a one-size-fits-all approach, ignoring individual student differences. However, the application of Internet technology allows education to better meet learners' personalized needs. Using AI and big data analysis, educational platforms can better understand learners' subject interests, learning styles, and progress, offering customized learning paths. This personalized education model helps improve learning outcomes, stimulates students' interest in learning, and fosters their ability to learn independently.

Furthermore, the Internet has transformed the dissemination of education. Traditional education mainly relies on face-to-face teaching, while the Internet liberates educational dissemination from temporal and spatial constraints. Through online videos, live classes, and blogs, educational content can be delivered in real-time, allowing learners to access knowledge anytime, anywhere. This flexibility and convenience adapt education to the fast-paced lifestyle of modern society.

Clearly, the Internet's transformation of education is a revolutionary change, providing learners with broader, more autonomous, and personalized learning opportunities. Through globalized educational resources, the emergence of online learning platforms, the promotion of personalized education models, and changes in educational dissemination methods, the Internet has not only altered how learners acquire knowledge but also shaped a more open, flexible, and innovative educational ecosystem.

However, despite the many positive changes the Internet has brought to education, "Internet Plus Education" is not perfect and has its limitations, which can be manifested in three main aspects:

### ➤ ATMOSPHERE

Educational attainment is particularly sensitive to the surrounding environment. People learn much better in a conducive atmosphere, without which the learning materials make little difference. If the majority of a class is keen to learn, those who are less keen could be motivated by the atmosphere and may gain a taste for learning as well. On the contrary, a good student could lose heart if placed in a class where nobody pays attention.

Existing online learning services still cannot provide a learning atmosphere that keeps people eager to learn and achieve results. Many pay the fee to start a course, but very few complete it. Learning can be tedious and exhausting for many, and it is hard to sit in front of a computer alone in a room to receive an education. If no change is made to the learning model, it is very unlikely this business model will be a success.

### ➤ MOTIVATION AND PRESSURE

In our educational journey from childhood to adulthood, the pressure of learning inevitably increases as we grow, which is a normal scenario because without some level of pressure, it's challenging to achieve results. Traditional classroom education often comes with supervision and encouragement from homeroom and subject teachers. When students are not attentive in class, fail to complete homework, or achieve unsatisfactory grades, there are corresponding punitive and remedial measures. However, online education based on the Internet is entirely different. It lacks supervision and punishment, which can further encourage people's inherent laziness, potentially leading to a surge in dropout rates.

During the COVID-19 pandemic, due to quarantine policies, most schools had to implement online teaching. While this ensured the continuation of coursework, the actual effectiveness was significantly inferior to in-person instruction. In traditional classrooms, students and teachers can interact face-to-face and ask and answer questions directly. When classes shifted online, many students just woke up to mark attendance and then went back to sleep.

It's evident that online education relies on students' self-directed learning capabilities, which is precisely what many Chinese students lack. Even in foreign countries, where MOOCs are well-developed, the completion rate of students who truly finish their online courses is less than 5%.

#### › LACK OF PERSONALIZATION

Traditional classroom teaching in China is generally delivered to classes of 30 to 50 students. Students all have a tutor assigned to them throughout their school life—from younger grades through university. Every student learns in a different way and has their own individual needs. A personalized study strategy may be particularly important in achieving learning targets for some students.

Personalized education is the future of education, and this may be difficult to achieve with off-line learning. It should be pointed out that because much online education is only off-line lectures moved to the line, so much of the so-called online personalized teaching is often not as good as off-line education. For example, many educational institutions only record the teaching process of the off-line classroom and place it online for students to download.

## 10.2 CHANGES IN THE FUTURE EDUCATION

Personalized and effective education has only one place to go in the future, and that is to join forces with wearable solutions.

The education system, which includes the format of lessons and approach to learning, will have to radically shift its focus. The objective will no longer be on the class as a whole, but on individual learners. Knowledge will no longer be about achieving certain quantifiable targets, but about the use of knowledge in new ways to reach higher levels of understanding about complex issues.

### 10.2.1 Immersive Education

In the movie *Inception*, there are some interesting plot lines. There is a scene in the movie where Dom Cobb's wife is sitting on a windowsill, ready to jump in order to wake from her dream. (This is established in the story: a person in a dream will be woken by a fall.) Cobb explained repeatedly to her that they were in reality, not a dream, but she remained certain that it was a dream. She wanted Cobb to join her in the jump from the window ledge that would end the dream.

Why couldn't she tell the difference between reality and a dream? This is a profound question related to the human subconscious. Cobb planted the idea in his wife's subconscious that the world was a dream, although it wasn't. To deny something that you have personally experienced before is very difficult. The same is true in learning.

We know some people have "book smarts" and others "street smarts." Individuals in the latter category often win out as knowledge from real-life experiences often registers better in the mind. This is what immersive education is all about. It uses VR devices to create an experience-based learning environment and structure.

For example, when a medical student is learning the basics of anatomy, a VR device could help the student visualize the circulation system to help the student take in the necessary information to improve as a medic. A VR device could also take a history student to the past to attend Lincoln's Gettysburg Address instead of learning the speech on paper. VR technology allows information to run much deeper into the human mind.

Repeated immersive learning is the most promising direction for the future. It blurs the boundaries between false and real experiences. Simple knowledge memorization is camouflaged as a personal experience that could register deep in a person's brain. In physics, the laws of motion could be learned in a much more interesting way. Students could be put in Isaac Newton's place to experience the process of discovery. The key to learning is understanding, a level higher than knowledge memorization. True understanding is the essence of learning.

What business opportunities could this educational approach bring? The advance of VR technology is the key one. Existing VR devices are not yet optimized for education purposes. The user experience needs further improvement. Ideally, we would like to see VR devices lighter and easier to wear, with many education and learning packs developed and ready to use. Current educational content is not adapted for use with immersive education. The development of various VR versions of content packs holds great potential in terms of business opportunities. Those able to develop innovative and easy-to-use resource packs will succeed in the future education market.

### 10.2.2 Downloadable Learning

In *The Matrix* film trilogy, humans are connected to a matrix with tubes plugged into their brains. This matrix was controlled by a supercomputer ruling over humanity. A small group of humans learned the truth and returned to the real-world. Then they decided to fight back.

There is one scene in the movie where the resistance team needs to go back to the virtual world but needs more skills to defend themselves. The solution was to download this skill directly into the human brain—within seconds it is done and ready to use. In the popular Indian movie *PK*, the alien PK could learn any human language just by holding the hand of a person who speaks that language. These ideas seem totally fictional and unreal. But are they really that far off? Imagine the future where learning could be done by hacking the brain using a "plug-and-play" method. This would yield significant potential efficiencies.

Then, such an education can only be done with the help of wearable devices. How about wearable devices? What kind of wearable? A head-mounted wearable based on a BCI. It can be a wearable device such as a helmet, and then can be densely covered with a variety of contacts to contact the scalp, through some kind of similar to the neural flow of the way (said not very accurate, their own brain to make up), in the way that our brain produces memory to guide the knowledge into the human brain. At present, many companies have already begun related research.

Knowledge packs, such as the entire history of ancient China, could be downloaded onto the wearable device and transmitted into the brain. How it might feel during the process of hacking your brain is hard to know. You might be amazed, when you find yourself like PK, speaking a foreign language fluently as if you have always known it.

This learning method likewise presents great business opportunities in the development of resource packs. Both downloadable learning and immersive learning could co-exist in the future education market. The experience is different between the two. One is focused on the immediate acquisition of knowledge, and the other is on adding fun to the process of learning.

### 10.2.3 Metaverse Education

Metaverse education essentially refers to wearable education facilitated by VR technology. With the advent of the metaverse concept and its growing public understanding, a profound impact is being witnessed across various industries. In January 2024, Japan's Aomine Next company announced a metaverse curriculum named "Metaverse Students" for Yuushi Kokusai International High School, drawing widespread attention in the educational sector.

Yuushi Kokusai International High School, a correspondence school with no geographical or time constraints, is renowned for its open enrollment policy for everyone. The introduction of this metaverse curriculum takes these features to a new level. Students are no longer confined to traditional classrooms but attend classes in the metaverse as virtual avatars. This innovative learning method not only breaks the barriers of location and time but also provides students with a more immersive and interactive learning experience.

In this metaverse curriculum, students will use VR devices for free and engage in course learning within the metaverse using tools like Zoom. The VR platform "Planeta" serves as both a communication tool for students and teaches them how to create metaverse and VR spaces. This means students not only acquire knowledge but also master critical skills for the future metaverse field.

Furthermore, the school plans to conduct group discussions and speech guidance in the metaverse space based on students' career paths to meet their individual needs and enhance their professional skills and teamwork abilities.

Various methods are implemented to promote student interaction, including the use of the SNS tool "key" to establish a tight social network in the metaverse.

To enrich the metaverse curriculum, the school has planned annual events. The cultural festival "Yushi Festival" will be organized and held in the metaverse space, giving students ample opportunities to unleash their creativity in planning and organizing various activities. Additionally, the school will host an e-sports tournament to foster students' online planning, execution, and collaboration skills.

The school's learning software, "you-net DX," plays a crucial role in classroom instruction. Besides "live online classes," students can watch "on-demand classes" based on their interests. This flexible learning approach allows students to study at their own pace and according to their interests, enhancing learning outcomes. Even in live online classes, students need not reveal their real faces; they can attend classes as avatars, protecting their privacy.

Students are provided with free avatars to choose from and can also purchase and use paid items externally. Although students can opt for avatars of a different gender from their own, the school recommends choosing humanoid avatars as they are intended as communication tools. This open and inclusive approach enables students to freely express themselves and maintain harmony and stability in the metaverse community.

For parents, the primary concern is their children's learning outcomes and future development. Yuushi Kokusai International High School's metaverse curriculum not only offers high-quality educational resources but also opens doors for students to venture into the future metaverse field. Here, students can obtain a high school diploma and acquire essential skills and experiences needed for future society.

The Metaverse Students program is set to officially launch in April, and the school is currently organizing online orientation sessions for students and parents. This innovative initiative undoubtedly injects new vitality into the education sector and fills us with anticipation for the future development of education. In summary, the "Metaverse Students" curriculum designed by Aomine Next for Yuushi Kokusai International High School is a groundbreaking educational innovation. It transcends traditional education's limits and provides students with a broader and more diverse learning space. Here, students can freely explore knowledge, make friends, and plan their futures. As wearable VR technology advances, metaverse-based education is poised to become a new direction in the educational field, leading us toward a brighter future.

CHAPTER 11

# WEARABLE + GAMING

Since ancient times, games have been an important part of human existence. Plato saw games as conscious mimicking developed by all youth (be it in the animal kingdom or among humans) to help them learn new skills and meet the needs of living. Aristotle took a more hedonistic view of games, seeing them as a type of rest and entertainment with no clear purpose, to be undertaken after the work of the day is over. More recently Raph Koster, the Chief Creative Officer of Sony Online Entertainment, defined them in the following way: "Games are the art of math. They are teaching us the system of themselves."*

Computer games are a digital extension of a tradition that extends back through the ages. Looking back, video games became popular in the 1970s, with the first commercially successful game, *Pong*, launched by Atari in 1972 in an arcade format. The first home gaming console was launched the same year, and by the late 1970s, home gaming had become popular. Gaming has gone through a number of changes, from the simple pleasures of shooting down aliens in *Space Invaders* to shooting a round of golf on the Wii to recently introduced VR gaming.

* "Abstract Games with Raph Koster," https://www.raphkoster.com/games/interviews-and-panels/abstract-games/.

The future of gaming is clear: it will be an integral part of our lives, both at home and at work, in education, and in our fitness routines. Wearable devices will promote the gaming industry further forward, leading people into an unprecedented gaming world.

## 11.1 CHANGES IN GAMING

After decades of evolution, gaming has become a second home to many. Game developers have seized the opportunities provided by emerging technologies to generate groundbreaking video games that transport people from their reality to a different world. Improvements in computing power, sound, and visual displays have all helped in this. Using joysticks, mice, keyboards, or touchscreens, we can interact with new virtual worlds from the platform games of *Super Mario* to the immersive experiences of *Robo Recall* on the Oculus Rift.

Web games are currently one the most popular types of video games. First appearing in Germany, these browser-based games do not require users to download software and can be accessed from any location. They are popular among office workers, as they have a short loading time and are easy to close. There are also no computer memory or configuration requirements.

The gaming industry has been continually evolving and adapting. The main gaming platforms have shifted from TVs to Gameboys, PCs, and now smartphones. The gaming experience has improved through these different media.

However, no matter how the games evolve and update, they can never detach from the reality that serves as their foundation. Although games construct a virtual world, people cannot truly enter this world. The reason lies in the fact that, on one hand, games can create an engaging virtual world, but this is merely a sensory experience. The scenes in VR may appear lifelike, but in reality, people's bodies are still confined to physical space. No matter how realistic the game is, players need to interact with the virtual world through screens or other input devices, and the real physical sensations cannot be fully integrated into the virtual environment.

On the other hand, human senses and physiological functions are deeply rooted in the real-world, and games cannot alter this fact. Whether it's touch, hearing, smell, or other senses, these are crucial ways through which people gather information in reality, and games can only simulate these sensations through limited means such as audiovisuals. Although VR technology has made some progress, it still cannot fully mimic the diverse sensory experiences of the real-world.

The advent of wearable devices introduces new possibilities to address these issues. For instance, wearable devices can interact more directly with users' physiological indicators, such as monitoring heart rate and movement trajectories, making the game more aligned with the player's physical state. This adds more realistic and personalized elements to the gaming experience, bringing games closer to real-life. Wearable devices also expand the development horizons for technologies like AR and VR. With AR technology, virtual elements can blend with the real environment, enriching the gaming scenes. VR technology, on the other hand, immerses players deeper into the virtual world, providing a more vivid gaming experience. For the gaming industry, wearable devices represent a novel endeavor that will bring unprecedented transformation to the sector:

### 11.1.1 Changing the User Interface

The first change brought by wearables will be a change in gaming interfaces, with the replacement of keyboards and mice by the players themselves. Keyboards and mice have long been used as the major tools to connect users with the gaming device. However, no matter how well-made they are, these items still have a time lag between player action and the game system's reaction.

In fact, the best interface is no interface at all. Gamers perform best with their own bodies in reaction to the gaming situation. Wearable technology suits this perfectly as wearables are great at recording movement data. Gamers feel fully integrated into the gaming environment. Recorded movement data could be translated into gaming data, to be added to the gaming scenario. On the other hand, users could be playing games while walking and running. It may be possible to create non-gaming environment games with wearables.

### 11.1.2 Gaming Formats

The formats of games are going to see changes too. Games will evolve from a virtual online world into AR—a combined world of virtual and real-world elements. Online and off-line activities will be merged into one.

On the one hand, games are going to be part of our lives in all aspects through wearables. Traditional video games were, and are, largely separated from reality. Gamers have a real-life identity and a gaming persona. These two normally do not overlap. But wearables are going to change this. If you played a sports-related game recently, the smart shoes you wear would record your gaming movements and acknowledge that as exercise. Your daily jog could also be added to your online gaming account, giving you bonus points in your games. We will see the merger of online and off-line activities. The smartwatches and smart bands we wear today have also incorporated this idea, but the benefits of gamification have not been totally realized. The manufacturers of wearables have gone some way toward the use of gaming social networks based on physical activity, but this has not been fully incorporated into other games.

On the other hand, the use of VR headsets in gaming provides an immersive experience that blurs the boundary between the real and virtual worlds. This is another way of seeing the two worlds come together.

In the future, games will be more experience-based. With the increase of different gaming formats and the growth in the entire supply chain of wearable technology, this could very likely be the case. In recent times, we have seen outdoor games, indoor interactive games, and family TV games increase in popularity. Nothing offers a better interactive experience than using wearable devices in gaming.

Gaming control is one of the key elements that determines the development of gaming formats. If this area is addressed well, perhaps television-based gaming could take off once again. To achieve this breakthrough, video game developers need to work together with wearable manufacturers. For game developers, it is important to involve wearable developers early in the game design process, in order to achieve the experience they want to provide. They may want to choose a specific gaming type and engage heavily in a form of "co-creation." It would be very exciting to see new gaming experiences on offer that meet the users' needs and exceed their expectations with fully supported accessories and gaming products. VR-enabled television-based video games, supported by wearable-based controllers, will provide entertainment in ways not currently imagined.

### 11.1.3 The Games

The third change in the gaming industry will occur in the design of games themselves. Games of the future could be seen as a way to promote health and provide positive experiences rather than the overall negative image that gaming has today. The secret of such a change lies in the integration of the gaming industry with other industries. Gaming will become part of fitness routines, medical treatments, and many other aspects of life. The gamification of life means the games will be an integral part of daily life too.

The integration of game-based technology into health promotion will be greatly aided by the use of wearables. Gaming's movement away from the screen to AR landscapes offers great potential benefits to players' health and well-being.

Scanning the horizon for future possibilities, we can see the potential for gaming to integrate with many other industries, including fast-moving consumer goods, education, communication, IT, finance, and more. In the case of education, for example, it may be possible to use a mind-controlled headset device to combine learning with gaming, making a game out of a problem-solving lesson. Gaming could be used as an additional way of learning rather than a distraction. Gaming could be defined by healthy, fun experiences and improved satisfaction with life. Gaming could become integral to daily life—drawing on wearables to allow the game to go to places never before imagined. It may become a new "Game of Life." Everything you do could provide data to the game, allowing everything to be gamified and part of the fun.

## 11.2 WEARABLES AS GAMING PERIPHERALS

Among all wearables, VR devices are most favored by the gaming industry, already finding a warm reception in the gaming market. VR device makers also realize that the gaming industry, among all other industries, has relatively lower barriers to entry. Now let us look at the details of wearables currently used as gaming peripherals.

### 11.2.1 Meta Quest 3

The latest VR headset from Meta, Quest 3, boasts numerous advantages. It features an updated and faster Snapdragon XR2 Gen 2 processor, enhanced graphics capabilities, a higher resolution display, improved lenses, and redesigned controllers that enable MR functionalities. Utilizing pass-through color cameras, it merges the virtual and real worlds, similar to Apple's Vision Pro headset. Like its predecessor, Quest 3 supports gaming, creative, and productivity apps and offers surprising fitness applications. It can also connect to a computer to serve as a gaming headset.

### 11.2.2 Vision Pro

On January 19, 2024, the much-anticipated Apple's first MR headset, Vision Pro, was finally released. As introduced in Apple's previous release event, Vision Pro offers innovative interaction methods, such as a virtual keyboard, eye tracking, and voice recognition, primarily for work and entertainment scenarios. On the hardware front, Vision Pro's core processor and sensor chips are Apple's proprietary developments, which clearly aid the company in better controlling product design. Gaming is a major function set for Vision Pro. Apple claims that users can play over 200 Apple Arcade games on this device, including *NBA 2K24 Arcade Edition* and *Sonic Dream Team.*

### 11.2.3 PlayStation VR 2

PlayStation VR 2 is a high-end gaming console VR headset, featuring an HDR OLED display, superior graphics quality, built-in eye tracking functionality, and advanced controllers that deliver the best gaming experience. It now boasts exclusive titles like *Gran Turismo 7, Resident Evil Village*, and *Horizon: Call of the Mountain.* To date, PSVR 2 lacks any social metaverse-type software, seeming more like a headset designed specifically for launching and playing VR games. Many of its games are ports available on devices like Quest 2. However, as more games optimized for this hardware are released, PSVR 2 is expected to distinguish itself from other stand-alone VR packages soon.

## 11.3 VR GAMING EXPLOSION

In the movie *Her*, the lead character, Theodore, is a lonely man living in the future. Before he met and fell in love with Samantha (an AI system), he spent most of his evenings playing video games. In the scene, Theodore controlled a little character with his body. He was not using any VR or movement-based devices. It was an immersive 3D video game using projectors and glass wall displays. Although this is a science fiction scenario, it could be the ultimate version of what we call VR technology.

Today, numerous VR games have entered the gaming market:

(1) *Beat Saber*—once awarded the Best VR Game of the Year, remains popular among players even four years after its release. In this game, players slash blocks with lightsabers in time with the music, which is a great test of reflexes.
(2) *No Man's Sky*—a nominee for the Best VR Game Award, faced criticism for its extensive monetization, being labeled as one of the biggest scams in the video game industry. However, continuous improvements by the development team have gradually restored its reputation, and it has garnered an increasing number of supporters, marking its place as an outstanding VR game.
(3) *Half-Life: Alyx*—released in March 2022, immerses players into a role as humanity's only hope against an alien coalition, offering an engaging gaming experience.

Beyond VR, AR games have sparked a new trend. *Pokémon Go*, developed by Niantic, achieved global success, spearheading the AR gaming movement. Since its launch, the game quickly became a worldwide sensation. As of May

2018, it had over 147 million monthly active users, and by early 2019, download numbers exceeded one billion. By 2020, its revenue had surpassed USD 6 billion.

*Pokémon GO* uniquely blends the real and virtual worlds, offering a location-based AR experience. Pokémon are scattered across the real-world, prompting players to move around to capture them. When encountering a Pokémon, it appears in AR mode as if existing in the real-world. Players can also engage in Pokémon battles, which are likewise set in real-world locations (Pokémon arenas).

*Pokémon GO*'s success extends beyond gaming to its advertising model. The game's distribution of Pokémon across various real-world locations can attract people to specific sites. For instance, in 2016, a partnership with McDonald's Japan transformed their outlets into Pokémon arenas, increasing average daily customers by 2,000 per store. Subsequently, Sprint in the US collaborated with Niantic, promoting over 10,500 retail stores similarly. Recently, Niantic's new game, *Harry Potter: Wizards Unite*, teamed up with AT&T, turning 10,000 of its retail outlets into inns and fortresses in the game to draw in customers.

AR games can also integrate with indoor home environments, like *Nintendo's Mario Kart Live: Home Circuit*. Players use a physical toy car with a camera for racing, set up tracks at home, and AR superimposes traditional Mario Kart game elements. Only the race cars and furniture are real; everything else is graphically added through AR.

Leveraging Huawei's HMS Core AREngine, Huawei has collaborated with numerous Chinese Internet entertainment partners (including Tencent, NetEase, Perfect World, Miniwan, and others) to develop well-known games, fostering innovative gaming experiences and AR ecosystem development in China. For example, in the *X-Boom* game, players shoot at AR animal characters superimposed onto the real-world.

In the future, as wearable technology and related headsets mature, VR gaming will continue to evolve and expand.

## 11.4 GAMING WEARABLES ENABLE ULTIMATE INTERACTION

Mark Zuckerberg wrote about VR in this way: it's just like hanging out with your real friends anytime you want. He was quoted saying, "Imagine ... studying in a classroom of students and teachers all over the world or consulting with a doctor face-to-face—just by putting on goggles in your home ... By feeling truly present, you can share unbounded spaces and experiences with the people in your life. Imagine sharing not just moments with your friends online, but entire experiences and adventures."*

In traditional gaming, once the user leaves the screen, or puts down the handset, or exits the app, this pauses or ends the game. The same may not be true in the case of a game played on wearables, as the very means of interacting with devices are redefined. Anytime, anywhere, you can still have real-time interactions through a wearable device.

For example, Google Glass could be used as the scope for shooting games. You can have a fun game in the background of a real scene with Google Glass any time you like. Or you could discover virtual treasure hidden in physical locations. Wearable games can also be combined with exercise. In the 1990s, Nintendo launched a Pikachu pedometer, which could turn real exercise and walking data into bonus points that could be exchanged for additional gaming content. Later this idea was further developed by Nintendo with the launch of *Pokémon Go*. With mobile and wearable devices benefiting from better sensors, this function has become more powerful and interesting.

In terms of gaming functions, smart bands, and smartwatches obviously have more potential than Google Glass. Just imagine the monitoring functions that can be used to calculate your calorie burn during exercise, and the translation of that data into character elements within an MMO game. How cool would that be! There are of course much easier interactions with smart bands and smartwatches, such as the pedometer function. In fact, quite a few games have

* "Mark Zuckerberg: Here's Why I Just Spent USD 2 Billion on a Virtual-Reality Company," http://www.businessinsider.com/zuckerberg-why-facebook-bought-oculus-2014-3?IR=T.

used this functionality in gaming already, including *Walkr—Galaxy Adventure in Your Pocket*. In-app purchases offer another revenue stream, though it is possible to use the app on a free basis. As long as you are willing to walk, you can redeem those steps for points in the game.

Other than these, there could be more potential gaming-related functions for smart bands and smartwatches. For example, using vibrations to give us a thrill during intense gaming, or alarms to remind us to collect our daily rewards. They could also be used as independent somatosensory devices for gaming controls. These are not science fiction at all. Many companies have already started working on such applications of this smart wearable technology.

Wearables could make gaming more diverse. Above all, they would make the gaming experience more realistic, to the extent that it becomes part of normal life instead of a separate activity. This could make a positive impact on life too.

This current age is one that has been dominated by smartphones. We have already seen how much we have changed because of that technology. In the coming age of wearables, there will be much to look forward to.

CHAPTER 12

# WEARABLE + FITNESS

Fitness has never been more popular than in the past few decades. Yet, for many, it is a struggle to get out of their seats and start exercising. Laziness or not valuing fitness enough can mean people don't exercise as much as they should—and this certainly can reduce the numbers that meet the recommended national guidelines for physical activity. For those who do put in the effort, there may still be barriers to getting the best outcomes—possibly because of a lack of knowledge about the best techniques for optimal results. Boredom, lack of support, and a lack of good coaching can all contribute to a loss of enthusiasm necessary for regular exercise.

It seems that it requires an awful lot of effort to stick to a training regime. It is not a bad thing for businesses that consumers face significant challenges, as it means there are opportunities to exploit. But when there is no perceived need on the part of the consumer, what can a business do? Consumers may not be fully aware of their needs. These are what some might call latent or potential needs. Apple presents a classic example of spotting potential needs and developing products to address those needs.

It is similar to the fitness industry. Gyms are largely similar in their offerings and facilities. Apart from their physical locations, they are largely homogeneous. If you have a location with some equipment and a few fitness coaches, you can open a gym. And yet, fitness is still a subject of much discussion, which indicates some issues have not yet been fully addressed. The top reason behind such unmet needs comes from the fact that fitness and life are not well integrated. This raises the critical problem of how a business can help combine these two factors. Wearables certainly provide one solution.

## 12.1 FOUR FITNESS INDUSTRY TRENDS BROUGHT BY WEARABLES

Sports health is one of the areas that has really benefited from wearable technology, with many devices capable of monitoring an athlete's output, activity, heart rate, performance, environmental conditions, and potential health risks. Fitness trackers that can monitor things like diet, exercise, sleep, and activity have become hugely popular over the past few years, leading to steady growth in the wearable technology market. Many fitness wearables have evolved from simple fitness trackers into health monitors. The concept of using wearables to track health conditions is more accepted now. The next key issue after data tracking and recording is data mining.

Contradie's firm applies computational biology models to data collected from wearable devices to provide insights into what a user's body is doing and what it is trying to tell them.

Other companies—including Fitbit, Misfit (now part of Fossil), and Intel's Basis—are also doing similar things, while Google Fit, Microsoft Health, and Apple Health services promise to provide universal storage for the collected personal health data and a deeper, more meaningful analysis. Wearables will no longer be simple trackers but will provide solutions to a number of health issues.

### 12.1.1 Real-Time Health Data Tracking

The biggest difference between wearable-based training and traditional fitness training is the fact that wearables enable 24/7 tracking. The devices are generally lightweight, fashionable, and easy to carry. They monitor a user's heart rate, steps, sleep quality, body temperature, breathing, posture, and other body data. This data helps understand the heart's condition during exercise and recovery time after exercise in order to plan more effective and suitable exercise routines, making fitness training more targeted and useful.

### 12.1.2 Supporting Professional Sports Training

Wearables are particularly powerful in some areas of professional sports, particularly athlete training. They record crucial training data, which could be used to prevent injuries and generate better training results.

For example, there is a wearable device called the Motus sleeve, designed for those playing baseball or softball. It uses a lightweight sensor inside a compression sleeve to collect biomechanical data with every throw. The athlete can use this device to correct his or her technique, and this may help avoid injury to the ulnar collateral ligament (UCL). Combining accelerometer data with a gyroscope allows tracking of how the throwing arm moves, all of which can be downloaded to a smartphone and uses Bluetooth. Based on the arm movements and basic biomechanics, the Motus App is able to calculate the elbow torque on the UCL. It can track dozens of pitching and batting metrics, including hand speed, workload, power generated through the hips during a swing, and elbow torque.

This type of wearable is much more refined in terms of tracking and can command higher prices as a result. They are only used by professional athletes. Wearable makers may be able to produce these products for ordinary consumers who are amateur players of sports like baseball, football, basketball, or rugby. With higher performance and lower price tags, there is potential to position these products in different markets in the future.

### 12.1.3 Emotion Trackers

As wearables are closely attached to the bodies of users, they are the smart devices that have the best knowledge of users through precise data monitoring. We know that emotions can affect the physical condition of an individual.

Made by Sensoree, the Mood Sweater is another product for emotion tracking. People can know our mood without asking. The sensors on the product can transmit emotional data to the collar of the Mood Sweater, with the collar changing color according to whatever mood is detected.

British Airways also conducted testing on a Happiness Blanket. The blanket could monitor the user's mood through neural sensors on Bluetooth-connected earphones. The blanket emits a blue light to indicate happiness on the part of the user and a red light when the user is unhappy.

There are many similar products. I predict that these products could be most useful in reminding users to keep a check on their stress levels, so that people may take precautions and adjustments before more significant physical or mental health issues arise.

It is rather important to control stress for heart health and to maintain a healthy body fat level. Maximizing one's mental stability is a key to remaining in good overall physical health.

### 12.1.4 Burning Calories

Most people in fitness will understand the concept of calorie counting. It is very useful to keep track of the right amount of exercise to achieve good fitness levels. For those who do not have the habit of regular exercise, but want to develop such a habit, wearables could help improve self-discipline. Such users are keen to know how many calories are burned as a result of exercise. Most wearable devices sold today are not accurate in the way they estimate energy use, therefore the data results are not necessarily reliable. Existing devices currently use movement acceleration to determine the exercise type, then combine this with age, gender, body weight, and other data to calculate calories burned. This is obviously not an accurate method, nor does it help to calculate calorie intake from food.

Whether wearables will continue to be used by consumers in the future is determined by multiple factors—the benefits of around-the-clock tracking, the comfort involved in wearing the device, user-friendliness, data accuracy, and more. Users need to be assured that wearables can help them live healthier lives. Apart from well-developed activity trackers, more wearable formats will enter the fitness market. Business opportunities will be available for every part of the human body. We will see smart shoes, smart socks, smart knee pads, and smart wrist pads. There are promising marketing prospects for smart sports clothing, which could be worn by many.

## 12.2 SMART CLOTHES USED IN GYMS

Wearables are no longer novelties at the gym. People can be seen wearing chest straps, smartwatches, or smart bands to track their performance and hit particular targets. However, there are problems with existing products that are hard to ignore. One issue is awkwardness. For those who have never worn watches or bands, having an additional item on the wrist may not prove comfortable. To address that need, more smart garments have begun to appear on the market. Smart tops and tracksuits are as easy to wear as normal clothes, but they are also able to track heart rate, breathing, and other biometrics associated with physical activity.

OMsignal and Hexoskin are two major players in smart clothing.

OMsignal's smart T-shirt is embedded with multiple sensors capable of measuring the wearer's heart rate, breathing rate, breath volume, steps taken, activity intensity, heart rate variability, and calories burned. These data can be transmitted to a paired smartphone via Bluetooth through a "black box" clipped onto the shirt. This box continuously collects the wearer's data, regardless of its connection to the phone, with a battery life of two to three days. According to OMsignal, this T-shirt not only tracks the wearer's various data but also wicks away moisture, promotes blood circulation, and increases oxygen delivery to muscles. Moreover, the OMsignal app converts collected data into useful information, providing real-time feedback on the wearer's physical condition and stress levels. Additionally, it offers guidance and suggestions during workouts, allowing wearers to adjust their exercise plans based on their physical condition. Notably, the OMsignal smart T-shirt is machine washable.

In 2013, Hexoskin released its first washable smart garment, capable of capturing cardiac, respiratory, and physical activity metrics. Currently, Hexoskin's primary focus is developing innovative wearable sensors for health and mobile, distributed software for health data management and analysis. For instance, Hexoskin's latest product, ASTROSKIN, supports monitoring health metrics such as blood pressure, skin temperature, and blood oxygen levels. These health metrics have been clinically validated, which is crucial for preventing or providing early warnings for diseases like heart conditions and hypertension.

These smart clothes share one feature in common: concealed sensors implanted inside the garments. Smart clothing is not like other wearables, as the materials feel normal—there is no need for additional items to be worn that could get in the way. Using the technology without realizing it's there is in some ways the ultimate goal of the wearable technology industry.

## 12.3 FITNESS TRAINING FOR ALL: EXERCISING WHILE PLAYING

Bing Gordon, a partner at KPCB, the largest venture capital in the US, once said that every startup CEO should understand gamification.* No matter a startup's area of business, awareness of gamification during strategic planning is very important, especially when gaming is the new norm.

In healthcare, where medical treatment and routine care can prove tedious and monotonous, gamification can encourage users to participate more willingly. As a result, they can form healthier lifestyle habits and better utilize the health functions of the wearables.

The fact that gamification rewards people for a desired behavior is pure human psychology. Combining this element with medical treatment means patients are more likely to act willingly and enthusiastically to improve the management of their health conditions. The gamification of healthcare has incredible potential to benefit patients, businesses, and the wider society.

First, games release the desire to compete for fun. Once this desire is combined with wearable medical devices and their apps, it could form a community with high rates of retention. For example, when step or calorie data recorded by an activity tracker is shared through social media, it makes the data viewable by others. Once such information is shared, a sense of competition could start to appear among a circle of friends. An individual may want to take another short stroll if someone they know is slightly higher in their step count ranking.

* "Bing Gordon: Every Startup CEO Should Understand Gamification," https://techcrunch.com/2011/06/30/bing-gordon-every-startup-ceo-should-understand-gamification/.

As AI develops further in the future and wearables are able to read you better, they may even talk to you when they know you're receptive to messages like "That person has gone ahead of you." Then, you might stand up and figure out a plan to regain your place. Such competition would make daily exercise routines much easier to stick to.

Second, gamifying non-communicable disease management lets patients self-manage diseases in their daily lives. Ayogo incorporates games into Health-seeker, an app specially designed for diabetic patients and children prone to developing diabetes. Users may start by choosing a life target to complete, then win badges and medals by completing various tasks.

Any habit requires time to form, particularly habits for a healthy life, often requiring additional drive to deliberately repeat certain activities. Nannying, dictating orders, soft reminders, or punishments will not necessarily be effective. Gamification is different, however. Users form habits unconsciously through the playing experience. It is not only enabling but also entertaining.

Third, gamified health management could help collect better data for medical research. The role of social media in spreading popular games is significant. This could be used by researchers as a way of helping to study and track health management data from diabetic patients, for example. Researchers could look for communities that have health management games that are specially developed for diabetic patients. Data collected from these communities would have a larger coverage and be more complete. Apart from biometric data, there is a potential to study data from social media on the content of messages and analyze the role of the health service. It is possible to analyze the sentiment of messages on social media on certain topics, for example.

The clustering of patients in social media would not only be good for medical and pharmaceutical researchers for their targeted studies, but it could also be useful for patients to learn from each other and obtain peer-to-peer support. It may also be possible to market online consultations for patients with similar medical conditions using social media channels.

Of course, the main challenge facing gamification of health management is user stickiness. Although games could release the user's internal drive to continuously monitor their health and make necessary adjustments in life, the big question is how to retain users and make them loyal fans of the games they play. This can be tackled as follows:

(1) Upgrading
No good games remain consistently attractive without regular upgrades. Fast reiteration is the key to retaining user attention these days. Health management games need to learn this lesson in order to intervene effectively in users' daily lives.

Apart from making the games interesting in themselves, useful in-game databases and links to other resources are also important. These could draw on medical databases, tips from social networks, existing guides on health management methods, and so on. Users always value scientific, up-to-date, and effective health information.

(2) The social element
Incorporating social elements in games is common today. Most users prefer to play games with others. Competition also makes games more fun. Users also like to share experiences. Therefore, the social dimension of gamification is crucial and highly valuable.

As previously mentioned, Ayogo developed the software Healthseeker for diabetic patients and children at risk. As it is linked to Facebook, that platform provides a massive potential user base that could easily bond together as a supportive and competitive social network.

Social games also increase the sense of achievement during competition. While playing games, our bodies release dopamine, a chemical that stimulates the brain and makes you feel happy. Such feelings could then encourage users to carry on, forming a virtuous cycle and improved health.

(3) Strong and powerful incentives
When we talk about strong and powerful incentives, we talk about, for example, points collected in the game that can be redeemed for real-world prizes or cash. Mango Health is one of the schemes under which users can be rewarded with real

money when they achieve certain targets. However, health management games are different in some ways from other types of games. The aim of traditional games is fun, and rewards are directly usable in the game. However, health management games are trying to persuade users that it is worth employing this management tool for better health, encouraging user retention, and allowing for the collection of a sizable volume of data. Developers could then use the data to assist further commercialization.

UnitedHealthcare is a company out of Minnesota, in the US. Their *Baby Blocks* program is an interactive incentive program developed using smartphone technology to encourage regular visits to the doctor during pregnancy. There are over 50,000 pregnant women from over seven states participating in the program. These moms-to-be can unlock new game levels by making antenatal appointments. After completing some key appointments, they could receive gifts, including maternity clothes, baby accessories, and gift cards.

Other incentives include lower insurance premiums for those who exercise regularly, demonstrate healthier lifestyles, or show improved medical conditions. Free online and off-line consultation for users of a certain level is another incentive. Well-designed game levels and incentives could be the key to attracting and retaining users. Physical incentives are particularly important to drive users to unlock more levels by completing more tasks.

(4) Privacy protection

Data safety and privacy protection have become increasingly critical in the age of mobile Internet. The gamification of health management faces the same challenges. All involved parties—software and hardware developers, insurance companies, hospitals, and other medical service providers—could have access to users' private information. A patient with a non-communicable disease may be a willing participant in a gamified management scheme but still not want their personal information or even their condition to be shared. Certain diseases, including hepatitis or HIV/AIDS, have significant stigmas associated with them. Without careful design, the social interactive side of gamification could compromise the safety of personal information.

Apart from the risk of potential data leaks, participants need to work together with businesses to develop a user-centered data-sharing agreement. For example, if we want to develop a gamified health management app for diabetic patients, we want the app to monitor blood sugar every day to figure out the reasons behind elevated levels, so a healthier diet and exercise routine could be suggested. Users could also get points for completing these tasks. When the tasks are completed, they'd be presented with the choice of whether to share this with friends on social media and should be able to choose an option based on their privacy preferences.

Whether a device or software can flexibly and effectively protect privacy will be one of the core criteria in evaluating user experience in the future.

CHAPTER 13

# WEARABLE + ADVERTISEMENT

Like it or not, advertisements are everywhere. We can hardly avoid being influenced by them as we shop, so businesses are always looking for better ways of encouraging us to buy their goods and services.

What about wearable technologies? Have they been ignored because of the limits on screen size? By no means. Advertising agencies have been studying this opportunity for a long time. The key word in advertising is precision. Wearables provide the most precise information about the user, because they monitor the user around the clock in a more detailed way than any police stakeout could watch a criminal. Such information allows companies to develop more personalized real-time advertising with precise delivery.

## 13.1 WHO ARE THE MOVERS AND SHAKERS IN ADVERTISING WITH WEARABLES?

### 13.1.1 Wearable Advertising Engine

Tecsol, an Indian software company, introduced its advertising solution for wearables in 2014. They used the Motorola Moto360 smartwatch to display advertisements. Possible uses include letting the user know about a nearby coffee shop while walking on a street or, unprompted, telling the user the weather forecast as they head for an appointment on their calendar.*

Tecsol built a basic MVC model of a Cloud-hosted ad engine. A basic static advertising image can be uploaded and pushed to the wearable device. This pops up, and the user can click on it or dismiss it. The watch then sends data back for analysis.

### 13.1.2 Virtual Mock-Ups of Ads on Wearables

"Any device with a screen allows for an interesting opportunity,"† explains Atul Satija, vice president and head of revenue and operations at InMobi, a maker of mobile ad tools.

InMobi has a team of developers creating virtual mock-ups of ads on smartwatches, head-mounted displays, and other gadgets to get a feel for how they can serve as a platform for advertising agencies.

Millennial Media and Kiip have also joined the search for viable wearable-ad technology, underscoring the appeal of the devices as marketing platforms.

### 13.1.3 TapSense's Apple Watch Advertising Delivery System

Before the release of the Apple Watch, the mobile marketing company TapSense introduced an advertising delivery system specifically for Apple's device. This platform allows developers and merchants to place ads on the Apple Watch, featuring high localization and integration with Apple Pay.

The developers at TapSense believe that localization is a key feature of wrist-based advertising. Leveraging the GPS functionality of the iPhone, the connected Apple Watch can display ads based on the user's location. Integration with Apple Pay allows merchants to distribute coupons and other promotional items, enabling users to "use coupons with a swipe of Apple Pay." However, it's uncertain if Apple will permit TapSense to deploy ads on the Apple Watch, as

* "Quick, Wearables, Hide! The Ads Are Coming …," https://venturebeat.com/2014/07/16/quick-wearables-hide-the-ads-are-coming/.

† "Advisers Target Wearable Gadgets as Next Ad Frontier," http://www.inmobi.com/company/news/advertisers-target-wearable-gadgets-as-next-ad-frontier/.

TapSense has stated in their blog that their service is not yet able to integrate with Apple Pay.

Moreover, the mobile shopping company inMarket has announced plans to follow suit with ad pushes for the Apple Watch, allowing users to receive promotional content on their device while shopping, possibly using technology similar to iBeacon, although the exact technology they will adopt remains unclear.

## 13.2 ADVERTISING WITH WEARABLES

The business opportunities brought by wearables are based on the valuable, unique data they collect. After processing and analysis, more useful and detailed user information could enable marketers and advertising companies to push more precise ads to the consumer in a new and exciting way. This would take advertising to a whole different level in terms of targeting and personalization. This is significant for the future of mobile marketing.

### 13.2.1 Smart Glasses

> *Knowing where I am is interesting. Knowing what I'm looking at or studying for three to four minutes is more interesting.*
>
> —JULIE ASH, ANALYST AT FORRESTER RESEARCH INC.*

The use of eye tracking in smart glasses offers great potential for understanding the shopping habits of users—detecting what users are looking at, what catches their eye, and what they think about when they are shopping. The advantages of this advance in wearable tech are clear.

Google Glass has a patent for a gaze-tracking system, which follows a user's

* "Companies Hunt for the Next Ad Frontier," https://www.iol.co.za/business-report/companies/hunt-for-the-next-ad-frontier-1715734.

gaze to know the user's feelings. It can also generate a gazing log, tracking the identified items viewed by the user and the user's emotions at that moment in time.

Compared to other smart glasses, those developed by Google are apparently the most competitive in the advertising business, as they have the best user data support. Google Glass can provide information on nearby restaurants based on user preferences or tell the user where their friends are eating. Coupons or discounts can be offered based on big data analysis.

Google Glass also has a better human-machine interaction experience than most others. This makes Glass users more likely to accept ads on the device. There might be voice-controlled interactive ads and user-selected ads with hands-free wearables like Google Glass.

Smart glasses and smartwatches have the largest screens among all wearables, but the largest are still relatively small. If advertising companies try to fill the screen area with ads, they will not necessarily be well-received.

### 13.2.2 Smartwatches

No genuine smartwatch ads have hit the small screen yet, but many marketers are attracted to this idea. The Apple Watch has been visualized as a billboard on the wrist by many mobile advertisers. Limited by the size of the screen, ads on smartwatches will be very different from those on smartphones.

A smartwatch may look like just another screen to conquer after television, computers, and smartphones. But it is a new frontier between consumers and advertisers.

John Havens wrote in *Hacking H(app)iness*, "It [your smartwatch] might say, 'Your pulse just went up, lay off the coffee.'" Havens also foresees slightly more insidious use of smartwatches. Retailers, he speculated, could monitor your pulse as you walk through a store. If you see an item that makes it race, they could then present you with an offer.

Such an advertisement delivery format would redefine "preciseness" in advertising. Traditional precise advertising is based on big data analysis. By tracking your Internet search history, web pages pop up related ads. The advertiser may not truly know what you like or whether you bought the products or not.

The mind-reading ability of smartwatches, however, is in a different league. They could be used to measure user preferences through heart rate monitoring. Precise preference data could be generated through analysis of the accumulated data of multiple users over time.

Wearables are already collecting a great deal of biometrics. Through pulse tracking, smart wrist-mounted wearables have great potential in healthcare, but will they be fantastic for advertisers keen to understand consumers' feelings? It could be a great application of this technology, but in terms of whether it will be as useful in advertising as in healthcare, we will have to wait and see.

### 13.2.3 Mind-Reading at Your Fingertips

Wearable device maker Personal Neuro Devices has a tagline: "Your mind at your fingertips."* What does this mean? This indicates that future ads could be pushed to users as a result of reading their minds. When you feel low, wearables may push ads for chocolates or music albums you'd like. When you feel hungry, the device can read your brain and decide which cuisine to recommend—Chinese, Italian, or junk food, for example.

The potential for mind-reading via technology has been exhibited in a number of different studies in recent years. Functional magnetic resonance imaging technology has been used by researchers at Cornell University to help decode images in the mind. It is possible to work out what people are thinking about based on these scans of the brain. Similar studies in Japan have shown promise in helping people with locked-in syndrome communicate.†

I believe this must be thrilling news for advertisers, but for users, this may raise concerns. If brain scanning and mind-reading technology advance to the stage where private thoughts are immediately known, this would be an ethical and moral minefield. This new concept of marketing is called neuromarketing.‡ It is indeed a challenging concept.

---

* "Could Wearable Tech Read Minds to Sell Ads?," https://www.cnet.com/news/could-wearable-tech-read-minds-to-sell-ads/.

† "Device That Can Literally Read Your Mind Invented by Scientists," https://www.independent.co.uk/news/science/read-your-mind-brain-waves-thoughts-locked-in-syndrome-toyohashi-japan-a7687471.html.

‡ "Neuromarketing," https://en.wikipedia.org/wiki/Neuromarketing.

Wearable marketing may take on many forms in the future, but so far it seems more practical to focus on GPS-enabled advertising of local points of interest, with potential for targeted discounts and integrated payment through wearables. This is the easiest marketing approach for users. After all, nobody would be offended by a discount offer, would they?

## 13.3 CHALLENGES FACED BY WEARABLE + ADS

### 13.3.1 Advertising Media

Google predicted that future ads would be placed in more unexpected locations, on radiators and refrigerators in the home, on car dashboards, through glasses, and on watches. We can imagine using fridges or dashboards as advertising media, as they have relatively more space. However, wearables are not quite the same as these surfaces.

First of all, existing wearable devices all have very small screens, or none at all in the case of smart bands, smart rings or smart clothes. How can these be used for advertising?

An American startup released a smartwatch that is able to project information onto the back of the user's hand (fig. 13-1). With a built-in micro projector, it can display the time and other notices from the linked smartphone onto the

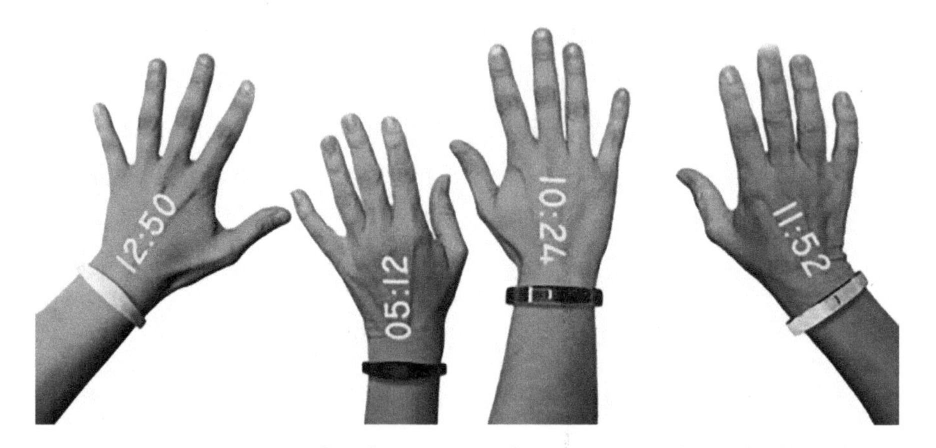


**Figure 13-1** Smartwatches that project information onto the back of hands

back of the hand. For small-screen or non-screen wearables, image projection could be a solution that provides space for revenue-raising advertisements.

However, this still doesn't solve the entire issue. The overall trend of future wearables is to be less visible, with the sensors more closely incorporated into human bodies. When this happens, there will be nowhere left for information projection. Where can the ads go?

The use of voice is one answer. The next stage of human-machine interaction will be voice control. Users will probably shift from the receiving side to the requesting side for ads as well. For example, if you want to buy new clothing, the hidden device detects your desire and scans available products in the background based on climate, temperature, user body shape, user preference, and price range. The filtered search results would then be pushed to the user to browse. How could this be displayed? Projecting such information in a 3D space with VR technology is what we can expect in the future (fig. 13-2).

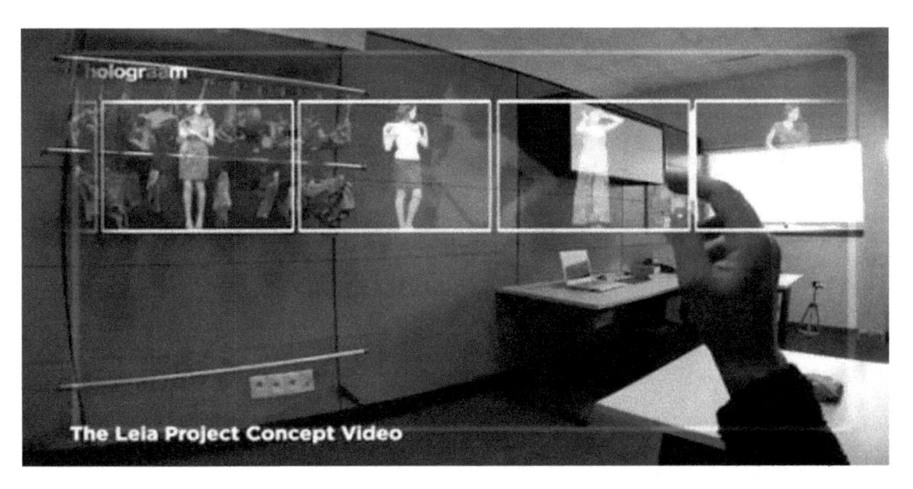


**Figure 13-2** Product display using VR technology

After voicing the request, "I'd like to buy some clothes," our own virtual images will soon be trying on different clothes in front of our eyes. Instead of viewing the clothing on models, seeing them on our virtual selves could make our shopping decisions easier and returns less likely.

Therefore, the ultimate display option for wearables might be based on VR technology, which is flexible and not restricted by any physical screens. We can view physical screens and projections as necessary steps in the march to VR, but these steps could take quite a long time, as core VR technology is a tough battlefield to conquer.

### 13.3.2 Consumer Attitudes toward Advertising

The modern concept of the advertisement was born in the late 1970s. Since then, the format, delivery, and media of advertising have all changed dramatically. Ads today find every possible way to present themselves before consumers, and yet the overall consumer attitude toward advertising has been going in the opposite direction.

OMD published a consumer report recently on attitudes toward advertising. In this report, it points out that consumers have extremely diversified attitudes toward mobile advertising. The majority (89%) have negative feelings toward mobile ads, and yet 75% also think the ads are fun. As many as 94% even say they believe ads are necessary.

Clearly, people have mixed feelings about advertising. In general, advertising is accepted, but we don't want to be bombarded with advertisements. So far, we don't have clear statistics on wearable ads, nor do we know what ad format users are more likely to accept with these devices. One thing for sure is that compared to traditional screens, wearables are more personal. Users are more likely to consider wearable ads more intrusive than ads on any other media.

As personalized and precise marketing has become the new normal for advertising agencies, some buffers remain between users and advertisers. Yet users still feel that their personal space has been invaded. Wearables bring a new relationship between consumers and advertisers. The devices will be responsible for displaying ads, setting ad preferences, and making ads more like functions that support a better quality of life for users and reflect their own values.

IDC's recent study shows that consumers are most welcoming toward products recommended through social media and their circle of friends. In other words, as long as the products are truly desirable, it doesn't matter so much if they reach them in ad form or not.

AI-enabled wearables are likely to further blur the boundary between marketing and life. The relationship between consumers and advertisers will also be redefined. Maybe one day, even the word "ad" will no longer be used in the traditional sense.

### 13.3.3 The Inevitable Conflict between Big Data and Privacy

The objectives of commerce and privacy are often in conflict, especially so in today's big data world. As data processing power continues to improve, whoever wants to make use of the collected data can figure out rather comprehensive personal profiles for any given user. In a way, everyone is being scrutinized by thousands of eyes, from all different angles, around the clock.

Users, on the other hand, enjoy all the convenience and personalized services made possible by big data but worry about violations of privacy at the same time. This anxiety was clear in the case of Google Glass. Even though the device was not guilty of anything, people were still worried.

People are sensitive to the risk of their privacy becoming compromised with the mobile Internet, and such fears only become worse in the case of wearable technology, as the core of the technology is to generate value from personal data. Wearables also present a different marketing platform. Advertising would be more invasive, making privacy issues an even bigger problem.

Capitalizing on big data means battles between users and businesses are going to be inevitable. Businesses want to seize the opportunities made possible by knowing more about their customers. Users are going to fight for both their privacy and the big data-enabled benefits they enjoy. In fact, this is an unsolvable conflict. The balance will constantly shift between the two sides. Advertisers need to know the true thoughts of consumers in order to deliver precise ads. When true thoughts cannot be openly requested, all sorts of "other means" come into play. Consumers' fears, on the other hand, largely stem from the unknown trajectory of this technology. No one can tell how far businesses are going to push the boundaries. This fear makes the privacy conflict a potentially drawn-out affair.

Finding the balance between big data capitalization and user privacy protection is one of the key issues that need to be addressed in the age of wearable technology. The "right to be forgotten," under EU law, allows users to delete information that could potentially violate their privacy. This is the first step the EU took to protect the privacy of the general public. It may not have significant impacts, but it is a reminder to all. The commercialization of big data is unstoppable. Privacy protection is also receiving more and more attention. In the future, individuals may be empowered by law to protect their privacy and enjoy additional rights.

With the radical change in format, value, and media of advertising in the age of wearables, the sky is the limit for potential value across the entire wearable industry.

CHAPTER 14

# WEARABLE + SMART HOMES

Today, smart technology is ubiquitous in our lives, particularly in home settings. Nearly every household item one can think of or see has undergone a smart upgrade. For example, light bulbs can be voice-controlled, robotic vacuum cleaners for cleaning are available, washing machines are fully automatic, and refrigerators have developed a variety of smart features.

In 2023, Siemens developed a refrigerator capable of detecting odors—the Siemens eNose smart refrigerator—precisely targeting consumers' health needs for food freshness detection.

Using our eyes and nose to judge whether food in the refrigerator has spoiled is an inexact science, filled with uncertainty. If one type of food in the refrigerator spoils, the bacteria it produces can spread to other foods, causing them to spoil as well. Therefore, timely and accurate detection of food spoilage is crucial.

So, if the spoilage of food in the refrigerator releases gases, could these be detected using a gas sensor? With this idea in mind and leveraging Bosch Sensortec's gas-sensing technology, Siemens successfully developed this smart refrigerator.

The sensors in the smart eNose continuously monitor the odors inside the refrigerator 24 hours a day. If it detects the emission of odors indicative of spoilage, it immediately sends a warning to the user's smartphone and automatically activates its deodorization and sterilization function, which has a 99.99% sterilization rate. With several advantages, including innovative food preservation technology, this refrigerator won the AWE 2023 A' Design Award.

## 14.1 MAN, THE CENTER OF THE IOT

Laziness is part of human nature. We naturally prefer enjoyment and activities that require less effort. That preference has also driven the development of new technology. When we examine the development of the mobile Internet, everything was anthropocentric—humans were at the center. Looking ahead to the IoT, it is even more so, making life better and easier. The first step is to build smarter homes that can interact with wearables.

Wearables shorten the distance between humans and devices, enabling control through novel interfaces in the time of AI. Mind control is moving from science fiction to reality, with the integration of digital and biological technology and the use of big data and cloud computing. Based on the brain control information database built through neuroimaging, brain implants could be placed into corresponding control locations to intercept or record (or stimulate) signals transmitted from brain neurons, in order to acquire specific information or execute orders given by the person with the brain implants. Using this technology, implants could be placed at different locations in the brain to achieve different purposes.

The man could then become the center of an AI terminal. The smart home would not need a computer as a control center, as the human brain would be an integral part of the system. Your thoughts would initiate instructions to control the appliances around you.

### 14.1.1 A 24/7 AI World

Science fiction films often present dramatic pictures, with a super brain in the future surrounded by AI screens everywhere.

Low-vibration smart alarm clocks might be able to set themselves half an hour before your ideal waking time, in combination with sleep tracking data, to ensure a good start to the day. Breakfast could be prepared totally by machines. Well-balanced, nutritious menu options could be displayed on a VR smart screen.

Before you head off to work, your smart device could plan your commute based on real-time traffic flows, drawing on big data and cloud computing. An optimal route would be planned with a nearby parking space pre-booked, to minimize waiting time.

Once you leave, the smart home can automatically enter an empty mode, with lighting and air conditioning switching to energy-saving settings. Some appliances could start preparatory work or take stock, so food supplies are ordered online for timely delivery. A cleaning robot could start its work around the home, and switch to stand-by mode once cleaning is completed. Before you finish work, you indicate your meal preference through a remote device to enable AI appliances at home to start preparing dinner.

When you arrive home, a smart device can open the garage door for you, and a biometric ID recognition security system will let you in. All ambient settings will be adjusted according to your preferences, including lighting, sound, AC, TV, or wall displays, creating the most comfortable home environment for you. If you fancy a party, there is no need to go out. All your smart home appliances will prepare a fantastic party night with delightful food and drinks as well as a suitable party setting. After a nice soak in the bath, you could greet your friends and family using VR video briefly and go to sleep. Your smart home could again adjust your environment to best ensure quality rest, with ongoing sleep monitoring throughout the night. Smart homes could provide emotional help as well as practical convenience. An interactive AI voice system could be your assistant and friend to help solve issues and provide suggestions.

The smart home and the Internet and Things will connect all vehicles, appliances, and devices. They will all be equipped with sensors to process information and respond to situations. There will be direct conversations between devices and people.

## 14.2 WEARABLES: THE BEST SMART HOME TERMINALS

At the moment, most smart homes are controlled through smartphones or tablet computers. The smart home business is in its infancy, and smart homes of the future could be connected through wearables. Real-time data from the human body could be central to the efficient operation of smart homes.

Haier, a pioneer in smart home appliances noted above, developed a smartwatch that can control air conditioning units, one of the first of its kind in the world. Users only need to issue simple voice commands to control the air conditioning through this watch, including switching the unit on and off, adjusting strength, controlling the temperature, and more. Compared to the more common smartphone app controls, this is easier to use. It is a step forward from the general "device + app" model to smart, durable goods.

Furthermore, with the development of wearable devices, an era where everything is connected is gradually emerging. Home doors can be unlocked with wearables, light bulbs can be turned on, and even cars can be controlled by wearables. The OPPO Watch4 Pro boasts such capabilities. On September 6, 2023, OPPO officially announced that its new product, the OPPO Watch4 Pro, would be the first to launch with the Ideal Sensory Car Key feature. It will also support car key functions for many brands, including BYD, Changan, and Airways, providing a seamless travel experience for users of OPPO phones and watches.

Specifically, once the OPPO Watch4 Pro is equipped with the watch car key function, car owners can unlock their vehicles effortlessly via Bluetooth by simply wearing the watch. The watch automatically unlocks the vehicle when approaching and locks it when walking away, significantly streamlining the process of getting in and out of the car. There's no need to worry about whether the car is locked after leaving. Additionally, users can check the remaining electric and fuel range on their watch, giving them a clear understanding of their vehicle's remaining range and allowing them to prepare in advance to avoid range anxiety. Facing hot or cold weather, car owners can remotely control the car's air conditioning with the OPPO Watch4 Pro, turning on the air conditioner or heater before departing so they can enjoy the most comfortable temperature as soon as they get in the car, without waiting for the air conditioning to kick in after starting the vehicle. Moreover, the OPPO Watch4 Pro's key function includes features like remote window control and car honking for finding the car, crafting a comfortable and smart driving experience.

The most revolutionary change in the integration of wearables with smart homes is the move toward smart appliances becoming less visible in daily life. They will still be physically present, but human-device interaction will be increasingly driven by processed data derived from smart wearables. Imagine, for example, going for a jog and returning home. Instead of needing to open the door, switch on the AC, and wait for the temperature to drop, all these actions will no longer be necessary, thanks to smart devices. The smart door would recognize you and open it, while the ambient environment of the home would be automatically adjusted according to your body data before you arrive. Hot water will also be ready to use in the bathroom. Ultimately, there would be little or no conscious interaction needed.

Both smart home appliances and wearables are smart technology, therefore it is inevitable that these two will eventually merge. As an extension of our bodies and intelligence, wearables embody the interaction between humans and things. In the context of smart homes, wearables could be the key to opening up the wonders of smart appliances. They significantly shorten the distance between users and devices, enabling "zero interaction" in some cases. We can expect to see existing smartphone- or tablet-based smart home appliances being totally replaced by wearable-controlled devices very soon.

## 14.3 THE ADVANTAGE OF WEARABLES FOR SMART HOMES

Currently, smart homes are still at an early stage of development and require people to directly instruct the device to operate in a particular manner. The simplification and portability of controlling terminals for smart homes are very important. So far, most smart homes are controlled by smartphones and tablets. If the controller is converted to wearables, the user experience will be completely transformed. I believe the strengths of wearables lie in the following aspects:

(1) Simpler operability and improved user-friendliness. Compared to smartphone-controlled smart home systems, the use of wearables will make smart homes easier to run. Measurements from the human body or the natural movements of the body (blinking, waving, and so on) could give control signals that trigger actions. This is much easier than using fingers to press buttons or flick through menus, and would significantly reduce the time spent interacting with devices. Of course, we can also use voice interaction systems with AI to help us manage these smart devices by simply saying what we want them to do.

(2) A 24/7 attachment to the body, with no limits on time or space. Just as smartphones are more portable compared to PCs, wearable smart devices are undoubtedly more convenient to carry compared to other mobile devices. No matter how much we love our phones, it's unlikely we'll sleep clutching them at night. However, watches, wristbands, or even smart pajamas made from intelligent textiles and other wearable devices can stay with us even as we sleep. Being able to sleep with them might not be an advantage per se, but the capacity for all-day wear offers many valuable applications, such as continuous health and medical monitoring of our physical well-being.

(3) Inherent data capability: facilitating the realization of Web 3.0. The inherent data capability of wearable devices, which almost become one with the human body, provides a powerful data function from the outset. Our biometric characteristics and behaviors are digitalized through wearable devices. Integrating these data with home system construction will significantly improve the home living experience, achieving personalized customization that truly meets users' needs. Moreover, the advent of personal virtual data sovereignty will contribute to the realization of Web 3.0.

(4) Full human-device interaction. Currently, smart home devices connect to each other but do not so readily connect to the human user. All technical discussions on computer bus, Wi-Fi, or radio-frequency technology are based on connections between hardware devices. However, the core function of a smart home is to allow smart products to serve the users. The effective link between humans and products determines the level of connected and controlled automation. Human biometric data could serve this purpose, both on the move or in sleep mode.

CHAPTER 15

# WEARABLE + PUBLIC ADMINISTRATION

The possibilities for the use of smart wearables in public event management are significant. There was a case of a woman living in Calgary, Canada, who used her Fitbit data in a personal injury case to demonstrate that her activity level dropped after an accident. It is important to note that the data had been processed and analyzed by an independent third party, Vivametrica, before submission to the court. This was the very first time in legal history that data from a personal fitness tracker was used in court as evidence.

This case allows us to foresee even more applications in time, from integrating personal data from wearables, especially in public administration, e.g., crime management, criminal investigations, city planning, and surveying public opinion. Wearable technology will become increasingly useful and result in cost savings for various government agencies.

Professor Alex Pentland, a pioneer of wearable technology cited by *Forbes* as one of the world's top seven experts on big data, explained some of the underlying opportunities. He said that wearable technology could play an immeasurable role in many areas, including health, finance, urban planning, crime prediction, and more, providing an image of a future driven by data.

## 15.1 IDENTITY VERIFICATION: THE KILLER APP OF WEARABLE DEVICES

We have seen more and more methods of identity verification, along with increasingly complicated safety measures. Many smart devices are able to quickly and safely extract human biometric data like fingerprints, heart rates, facial features, and others to verify the user's identity. Compared to traditional encryption methods, encrypted biometrics are far superior and will gradually take over as the mainstream identity verification method for social media, smart devices, and some payments.

Wearables are the ultimate solution to make this method absolutely safe and secure. Why is this? Wearables know you much better than other smart devices. The main function of the wearable is to collect data from the user. After being processed and passed back, the data can become a unique verification code. In other words, identity verification based on wearable devices could not only identify a person through individual biological features but also generate a unique and irreplaceable identity code that is derived from a set of specific and abstract data, including heart rate, blood pressure, lipids, facial features, skin features, personal preferences, etc. This is exactly why wearable technology is so incredibly fascinating. No other application fits as perfectly as this one.

We are living in an age when everyone is worried about privacy, identity theft, and fraud. Data safety has become a major source of concern for individuals and businesses alike. Nobody can be totally free from safety concerns while enjoying the conveniences of the mobile Internet. That is why we all have to remember so many passwords and PIN codes. In fact, if you have used Alipay, you would know that every link is interlocked. Even the virtual keyboard for typing in the passcode is encrypted. Users are incredibly concerned about safety.

### 15.1.1 Security Checks Based on Wearable Devices

How can identity verification, the killer application for wearable devices, best interact with public life? In my view, wearable devices will become the core to support all future public administration with their unique identity verification function. It will become so crucial that, without this, almost nothing will work in the future. Right now, we can already see the huge obstacles faced by various sectors due to the difficulty of identity verification.

In Shenzhen, one of China's Special Economic Zones, local officials report that security equipment just for the Shenzhen Metro costs RMB 120 million (USD 17 million) even before maintenance costs are considered. Each month, the labor cost is over RMB 5 million (USD 714,000), and the annual cost exceeds RMB 60 million (USD 8.5 million). When the depreciation of equipment is taken into account, the annual cost of security screenings, just for the Shenzhen Metro, is approximately RMB 100 million (USD 14 million). The total revenue of the Chinese Football Association Super League in 2013 was around RMB 220 million (USD 31.4 million), with a net profit of RMB 100 million. The annual cost for security at the Shenzhen Metro was the equivalent of the annual net profit of the Chinese Super League. This is just for the Metro. There are other public transport hubs, including airports, bus stations, railway stations, etc., all requiring various levels of security. These places all use traditional security checks, which are labor-intensive and time-consuming for all. The potential market for wearable-based security checks in China is enormous.

On March 25, 2014, Spring Airlines was the first to successfully board passengers wearing smartwatches using QR code boarding passes. Ticket verification, security checks, and boarding were all completed swiftly.

By placing your wrist close to the ticket-check machine, the process can be completed within seconds. The same is true for boarding a bus: there is no need to fumble in your bag anymore, looking for your public transit pass. Just a tap of the wrist is all you need. Using a smartwatch to take public transport is already popular in Beijing.

In public, the first important role for wearables will be in security checks for public transport. When such wearable-based identity verification is developed, it would bring incredible convenience to travel. No more transit cards for buses or subways, no more flashing your not-so-flattering photo ID to strangers. A wristband will get you through all these gates.

From public transport security checks to other scenarios, this technology will free us from all those membership and bank cards, once and for all. In the future, people will no longer need to carry dozens of cards in their wallets. All these cards can be installed on smartwatches. With a single word, the payment interface could pop up. Cash withdrawals or redemption of loyalty card points will no longer require passcodes. Everything can be done with a single tap. Even if the owner loses the wearable device, it would not be a problem as it would provide no use to anyone else. Without the unique body signature of the owner, the watch will not work at all.

Having a wearable device will likely make life much easier for the user and others. A New York studio envisioned how wearables could make public transport more convenient. They conceived a wristband they called Relay, which could access subway data and display the data to its users. For example, when you have to make a decision on whether to take a taxi or the subway, Relay can give you the information on the next train to arrive at the station nearest you.

After analyzing data, a wearable device can inform its user of the best option at any given location, along with the best solutions. It can process information quickly and offer practical solutions. In the process of building smart cities, such technology may become an effective way to increase public transport usage, by making the commute much easier and more pleasant.

This will only be the beginning of wearable technology in public services. We will soon witness wearable devices used in other areas of public administration. When a citizen's data and credit records are planted into wearable devices, there will no longer be any need for ID cards and the associated risk of losing them. We won't need tedious identity check procedures when passing through security, or the arguments over real-name systems. This will not only save public administration costs, but also increase efficiency and effectively prevent crime.

### 15.1.2 Real-Name Systems Enabled by Wearable Devices

When you appear online as yourself, would you dare to spread rumors or act irresponsibly? I bet you wouldn't. We have seen various tickets, cards, mobile accounts, and even mobile phones that are now set up under people's real names. How long will it be before the real-name system is applied to other areas? The online environment is particularly in need of regulation. Many have taken advantage of anonymous virtual identities to speak without thinking of the consequences. People abuse others online, rant, and spread rumors, creating much frustration for the "Internet cleaners." So far, there has not been any effective way to solve, or even improve, this situation. That is because external pressure only makes the online abusers behave even worse. A real-name system is probably the only solution that could work.

On March 16, 2013, Sina and Tencent both began to apply a real-name system for Weibo, requiring users to provide personal ID information to register. Setting one's front-end user name is voluntary, but back-end registration is based on real-name ID verification. Since then, users with no ID verification could only browse without the ability to blog or reblog. In 2015, the State Internet Information Office in China implemented rules for identity information management. Under the principle of "real-name in the back end, voluntary use of real-name on the front end," Weibo, Baidu Tieba, etc., all adopted real-name identification policies to reinforce online monitoring and administrative enforcement. Today, the real-name system even covers almost all Internet applications.

The discussion here will not go into detail about the pros or cons of a real-name identification policy, but we can be sure of one thing: through the implementation of a real-name identification policy, the labor and material costs will rise for online social platforms. Take Sina Weibo for example. After adopting the real-name identification policy, three types of costs have gone up. First, the complexity of the system increased, requiring the installation of more servers; second, the website is required to check users' ID information; third, the need for privacy protections increases as do the difficulties, leading to higher operational costs for the website.

There are two types of "citizen ID information checks" available for websites in China, on offer at different price points. One is for personal users, costing RMB 5 (USD 0.71) each time. The other is for corporate users (e.g., payment websites), charging annual fees and transaction costs cheaper than RMB 5. The annual fees are equivalent to RMB 0.5–1 (USD 0.07–0.14) per ID check on average. According to the statistics published by Sina in 2013, there were already over 500 million Weibo users. At a price of RMB 0.5–1 per check, the real-name identification market for Sina Weibo alone would be worth around RMB 250–500 million (USD 35–71 million).

Wearable technology will accelerate the implementation of real-name systems at full scale, with public security fully covered. As mentioned above, this function can be seen as the killer app of wearables, and the need for identity verification is a significant driver for the emergence of this technology. Presently, when users register their ID number with Weibo, back-office staff still need to manually verify the information to complete the registration. But wearables can help this process happen instantly. When we register with any online social platform in the future, the validated identity information could automatically be passed to the operator in less than a second. The authenticity of the information would be 100% guaranteed, as any identifier would only work with the wearable worn by the specific user. If the wearable were removed from the user, the identification function would be temporarily locked down. Currently, no other security measures can match this.

It is difficult to get full insight into the number due to corporate confidentiality, but we can tell from Sina's example that the real-name identification policy is an open sea with massive potential for exploitation in the future. On the one hand, it will restrict individuals in terms of speaking in a public space and feeling responsible for what they say, hence making for a more orderly online environment. On the other hand, the government will spare no effort to assist in implementing such an environment.

Especially in the metaverse era, each of us will have a digital avatar or a digital virtual identity, which will be cross-platform and universal, based on quantum encryption technology. This is clearly inseparable from the digital identity constructed by wearable devices, which possess uniqueness and confidentiality. The core to achieving uniqueness and confidentiality lies in the biometric technology behind wearable devices. By utilizing individual unique biological characteristics such as retina, pulse, heart rate, fingerprint, palm print, and facial features, this unique digital ID can be constructed.

## 15.2 CURBING CRIMINAL ACTIVITIES

According to media reports, police in Dubai started using Google Glass to identify stolen vehicles. This device has two apps, one of which allows users to record traffic-related offenses, and the other one which allows for the identification of stolen vehicles by examining license plates. Police forces in New York, Los Angeles, and Byron City are also engaged in pilots using Google Glass.

The case of George Floyd in Minneapolis, Minnesota, has attracted a great deal of public attention throughout the entire world. One of the latest of many negative encounters between white police officers and African Americans in the US, there has since been an international uproar after Floyd, an unarmed black man, was killed by a white police officer who suffocated him to death.

Over the past several years, it has become a requirement for law enforcement to wear body cameras to capture all encounters with the public. However, despite this, there have since been many instances of officers turning these cameras off before approaching others. There have also been some instances of footage being withheld from the public. Though body cameras were initially meant to put an end to racist or indecent encounters, such instances as those described above have remained a subject of controversy and scrutiny. Wearables are particularly useful in helping police investigate cases. For example, head-mounted polygraphs could be useful in detecting suspects' lies. We could install sensors in a helmet to detect changes in brain waves or nervous system responses when the suspect is presented with pictures of the crime scene or the victim. If the suspect lies, no matter how good their poker face is, the deception can be detected and called out. This would be genuine mind-reading. It is direct information acquisition rather than interpretation through micro-expressions or body gestures. Our thoughts would be more visible than ever with wearable technology.

If a suspect could be quickly identified among a group of ordinary people, this would also enable criminal cases to be solved much more quickly. In Brazil, the police force uses a type of smart glasses. They are able to scan 400 faces per second, from up to 50 yards away. The images are compared with photographs in police records on the basis of 46,000 biometric identification points. Once a match is found, the glasses can highlight a suspect with red lines, avoiding spot ID checks all together.

It's clear that different wearable devices, including smart helmets and smart glasses, will play an increasingly important role in assisting law enforcement. The use of wearables will improve investigation efficiency, while also deterring criminality, contributing to a safer and more peaceful society.

Wearable glasses integrated with AI recognition technology can effectively and timely prevent and intervene in criminal behavior based on a predictive system of facial, expression, and behavioral characteristics big data. This applies to both city surveillance cameras and smart glasses worn by patrol officers. However, this also sparks significant controversy over the infringement of citizens' privacy rights.

Clearly, in the digital and intelligent era, or as we enter a comprehensive era of smart wearables, when human behavioral characteristics are datafied, how to ensure our rights are not interfered with or dominated by algorithms, and how to protect our privacy from excessive invasion by algorithms and governments, will be a significant challenge faced by our society now and in the future.

## 15.3 YOUR CITY BUILT BY YOU

The impact of the environment on human health and the potential role of wearables is perhaps best illustrated by the case of David Fairley. One day in 1998, Fairley, a statistician by training, was walking down a busy street in San Francisco to pick up his son from preschool when he suddenly felt weak and dizzy. On the way to the hospital, he suffered from a heart attack.*

Fairley partly attributed his heart attack to air quality. After years of investigation, he confirmed previous findings by a researcher in London on the

* "Wearable Tech Helps You Live in the Moment," https://www.scientificamerican.com/article/wearable-tech-helps-you-live-in-the-moment/.

relationship between particles in the air and an increase in a city's death rate, particularly deaths linked to cardiovascular and respiratory system problems. "Even though there were, of course, other factors, I believe walking up that street might well have contributed [to my heart attack]," he said. "Ultrafine particles are so small that they're fairly unstable. They don't stick around. They agglomerate into bigger particles or else diffuse out. If you look at the gradient from roads, concentrations of ultrafine particles are really a lot higher on a busy street. So, it really could make a difference to move one or two streets over."

After many years, David Fairley came to the realization that if one cannot change the environment, he should change his own route, choosing a street with a lower volume of traffic.

In this anecdote, we can see the effects of environmental factors on human health. Even though he was aware of this, Fairley was still slow to act. How can wearable technology help in this case?

We are able to install wireless sensors into anti-pollution facial masks which could collect real-time data from the streets, e.g., particle concentrations in the air, carbon dioxide density, PM2.5 levels, etc. This can be compared with historical data from governments and academic institutions. It would then be able to send the correct information to the user in time to prevent any issues from happening.

If David Fairley had one of these devices, he would not have waited years before changing his walking route. What could have happened was his wearable device would have immediately sent him alerts warning that particle concentration levels on that street would pose potential health risks or even trigger heart attacks. It also could have provided recommendations for alternative routes. This would have given Fairley better information with which to make a proper decision on his route in time to avoid being exposed to adverse environmental factors.

### 15.3.1 Working Together with Governments and Academic Institutions

More and more wearable technology companies are providing data for academics and governments. When the air quality index reaches a certain level, the wearable device will warn its users, informing them of the exposure and potential health risks. With such information, air pollution control agencies could issue more rational standards and create more effective policies.

The main technical challenges lie in the analysis of big data and the development of standards, which are also massive business opportunities for many wearable technology companies or data analysis firms. The core of wearable technology is the collection and deep processing of data. Without analysis and feedback, data collection alone does not bring much value or revenue.

By providing analytic reports to government agencies, we could witness incredible benefits regarding the design of urban environments in the future. Accuracy and efficiency will be improved significantly in planning, policy making, and public engagement.

There are companies already exploring opportunities in this area. Jonathan Lansey is a data engineer at Quanttus, a wearable technology company.* This company is developing a new watch that can measure and analyze users' vital signs—including heart rate, blood pressure, and body temperature. This data can be used to assess how they respond under different conditions, including different air pollution levels or various weather conditions. The company is also seeking to partner with academic institutions, hoping to provide data for research studies. Lansey has said that Quanttus plans to provide user data to academic institutions. By using this data, hopefully, scientists can gain more insight into the impact of environmental factors on human health, and provide supporting evidence for government policy making.

"I think that's going to be a large branch of our contribution; we're building a business here, but there's an altruistic component, too," said Steve Jungmann, vice president of product management for Quanttus. "We're across the street from MIT, so we get to talk to people there with some fantastic ideas around biometric monitoring as attached to x, whether it be air quality, emotions, the ability to perform under pressure, any number of things," he said. The company is also making devices for various research purposes.

---

* "The Future of Wearables Makes Cool Gadgets Meaningful," https://www.theatlantic.com/technology/archive/2014/05/the-future-of-wearables-makes-cool-gadgets-meaningful/371849/.

### 15.3.2 Smart Cities

In June 2015 Larry Page, CEO of Google, published a post on Google+ explaining that the company would establish a new urban innovation company called Sidewalk Labs, to improve the lives of billions of people on the planet. Sidewalk Labs focuses on developing new products, platforms and collaborating partnerships, to solve problems in living, transportation, and energy usage. Google has a clear and simple company positioning: "people." The aim is to make people more comfortable in a city, freeing them from concerns about traffic congestion, parking availability, long queues for food, and poor air quality. How do we achieve these objectives? One solution is to make the city around us more intelligent.

IBM defines a smart city as a city that fully utilizes all available Internet information for better understanding and control of city operations, as well as the optimal use of limited resources. Technology will be the building blocks of future cities. Without technology, cities will go nowhere. As one of the forerunners for the smart cities of tomorrow, IBM has provided solutions for many countries and cities in terms of urban transportation management options.

By studying over 50 cities in developed and developing countries, IBM has found that all of the cities in the world have their own transportation issues. However, some cities have achieved significant results with the help of IBM Smart Transport solutions. Take Stockholm, for example. As the capital of Sweden, the city accommodates over 500,000 vehicles every day. In 2005, the average commuting time was 18% longer than the previous year. The Royal Swedish Academy of Sciences then began a collaboration with IBM to develop and implement a smart transportation system that was suitable for the area. In 2006, this smart transportation system was adopted, and in the following three years traffic congestion was reduced by 25%. Time spent waiting in traffic was also reduced by 50%. Taxi revenue increased by 10%, and urban pollution dropped by 15%. Stockholm also saw a rise of 40,000 new public transit passengers.

The smartness of a city cannot be reflected by traffic alone. Traffic is something on the surface of a city. Technology is what's underneath, and wearable technology can prove most useful in areas like healthcare and education. Wearables not only bring a hands-free experience but, more importantly, redefine our lifestyle. In the future, omnipresent mobile Internet will connect wearable devices to make distance learning, telemedicine, and taxi businesses all possible. With the rapid development of the IoT and smart cities, information flows will be more open and interactive in the future. Entire cities could be felt and understood on a different level. New government proposals, open for public comment, will be distributed to wearables through dedicated systems that welcome your comments and feedback. You could be participating in these matters before you know it.

You will be contributing to the planning and construction of your city, sharing issues and ideas with fellow citizens. At the same time, you can also be kept informed about the progress of each plan. At the moment, such privileges are granted through applications to certain government units with a lot of necessary paperwork and justification.

In the future, unified platforms may provide a space for sharing information and interaction between citizens, policy makers, and implementing authorities.

In cities of the future, all urban management will be established upon a massive and complete touchscreen-based interactive map. Urban managers only need to drag and click to complete the configurations. If you have watched the movie *The Hunger Games*, you can imagine what it may look like. Having solutions at your fingertips, no matter what your needs are, is part of what wearable technology can provide.

We can foresee that future cities will be digital twin cities built on smart wearables. Looking from a broader earth perspective, humanity will digitize the earth using smart wearables, ushering in a digital twin earth, which is the era of the metaverse as currently envisioned. In the era of the metaverse, the dual worlds of virtual and real will no longer be independent but will form a new state of seamless interconnection and interaction based on smart wearables and Starlink communication technology.

To build such an era of the metaverse and such a digital twin earth, everything fundamentally relies on wearable devices and the development and popularization of the smart wearable industry.

# INDEX

## ABOUT THE AUTHOR

KEVIN CHEN is a renowned science and technology writer and scholar. He was a visiting scholar at Columbia University, a postdoctoral scholar at the University of Cambridge, and an invited course professor at Peking University. He has served as a special commentator and columnist for the *People's Daily*, CCTV, China Business Network, SINA, NetEase, and many other media outlets. He has published monographs in numerous domains, including finance, science and technology, real estate, medical treatments, and industrial design. He currently lives in Hong Kong.